QUIRKOLOGY

Professor Richard Wiseman

QUIRKOLOGY

MACMILLAN

First published 2007 by Macmillan

This edition published 2007 by Macmillan
an imprint of Pan Macmillan Ltd
Pan Macmillan, 20 New Wharf Road, London N1 9RR
Basingstoke and Oxford
Associated companies throughout the world
www.panmacmillan.com

ISBN 978-0-230-70215-8

1 3 5 7 9 8 6 4 2

A CIP catalogue record for this book is available from
the British Library.

Printed and bound in Great Britain by
Mackays of Chatham plc, Chatham, Kent

Visit **www.panmacmillan.com** to read more about all our books
and to buy them. You will also find features, author interviews and
news of any author events, and you can sign up for e-newsletters
so that you're always first to hear about our new releases.

To Mum and Dad

What is the use of such a study? The criticism implied in this question has never bothered me, for any activity seems to me of value if it satisfies curiosity, stimulates ideas, and gives a new slant to our understanding of the social world.

Stanley Milgram, *The Individual in a Social World*

Contents

The mysterious Q-test

Before we begin, please take a few moments to complete the following exercise.

Using the first finger of your dominant hand, please trace out the capital letter 'Q' on your forehead.

There are two ways of completing the exercise.
Turn over the page to find out more.

You can draw the letter 'Q' with the tail of the Q towards your right eye like this:

In this case, *you* can read it, but someone facing you can't. Or you can draw it with the tail of the Q towards your left eye:

In this case, someone *facing you* can read it, but you can't. As we will discover later, the way in which you completed the task reveals a great deal about an important aspect of your life.

Introduction

What quirkology is, why it matters, and secret studies into the
science of tea-making, the power of prayer, the personality
of fruit, and the initiation of Mexican waves.

I have long been fascinated by the quirky side of human
behaviour.

As a psychology undergraduate I stood for hours in
London's King's Cross railway station looking for people
who had just met their partners off a train. The moment
they were locked in a passionate embrace, I would walk up
to them, trigger a hidden stopwatch in my pocket, and
ask, 'Excuse me, do you mind taking part in a psychology
experiment? How many seconds have passed since I just said
the words Excuse me . . . ?' My results revealed that people
massively underestimate the passing of time when they are in
love, or, as Einstein once said, 'Sit with a beautiful woman
for an hour and it seems like a minute, sit on a hot stove for
a minute and it seems like an hour – that's relativity.'

An interest in the more unusual aspects of psychology has
continued throughout my career. I am not the first academic
to be fascinated by this approach to examining behaviour.
Each generation of scientists has produced a small number of
researchers who have investigated the strange and unusual.

Maverick Victorian scientist Sir Francis Galton might be considered the founding father of the approach, and devoted much of his life to the study of offbeat topics.[1] He objectively determined whether his colleagues' lectures were boring by surreptitiously measuring the level of fidgeting in their audiences, and created a 'Beauty Map' of Britain by walking along the high streets of major cities with a punch counter in his pocket, secretly recording whether the people he passed were good, medium, or bad looking (London was rated the best, Aberdeen the worst).

Galton's work into the effectiveness of prayer was more controversial.[2] He hypothesized that if prayer really worked, then members of the clergy – who clearly prayed longer and harder than most – should have a longer life expectancy than others. His extensive analyses of hundreds of entries in biographical dictionaries revealed that the clergy actually tended to die *before* lawyers and doctors, thus forcing the deeply religious Galton to question the power of prayer.

Even the making of tea caught Galton's attention, when he spent months scientifically determining the best way to brew the perfect cup of tea. Having constructed a special thermometer that allowed him constantly to monitor the temperature of the water inside his teapot, after much rigorous testing Galton concluded that:

> . . . the tea was full bodied, full tasted, and in no way bitter or flat . . . when the water in the teapot had remained between 180° and 190° Fahrenheit, and had stood eight minutes on the leaves.[3]

Satisfied with the thoroughness of his investigation, Galton proudly declared, 'There is no other mystery in the teapot.'

On the surface, Galton's investigations into boredom, beauty, prayer, and tea-making may appear diverse. However, they are all excellent and early examples of an approach to investigating human behaviour that I have labelled 'quirkology'.

Put simply, quirkology uses scientific methods to study the more curious aspects of everyday life. This approach to psychology has been pioneered by a small number of researchers over the past hundred years, but has never been formally recognized within the social sciences. These researchers have followed in Galton's footsteps, and had the courage to explore the places where mainstream scientists fear to tread. Academics have:

- examined how many people it takes to start a Mexican wave in a football stadium[4]

- charted the upper limits of visual memory by having people try to accurately remember 10,000 photographs[5]

- identified the perceived personality characteristics of fruit (lemons are seen as dislikeable, onions as stupid, and mushrooms as social climbers)[6]

- secretly counted the number of people wearing their baseball caps the right way round or back to front[7]

- stood outside supermarkets with charity boxes quietly measuring how different types of requests for donations impacted upon the amount of money given (simply saying 'even a penny helps' almost doubled donations)[8]

- discovered that children's drawings of Santa Claus grow larger in the build-up to Christmas Day, and then shrink in size during January.[9]

For the past twenty years, I have carried out similarly strange investigations into human behaviour. I have examined the telltale signs that give away a liar, explored how our personalities are shaped by our month of birth, uncovered the secret science behind speed dating and personal ads, and investigated what a person's sense of humour reveals about the innermost workings of their mind. The work has involved secretly observing people as they go about their daily business, conducting unusual experiments in art exhibitions and music concerts, and even staging fake seances in allegedly haunted buildings. The studies have involved thousands of people all over the world.

This book details my adventures and experiments, and also pays homage to unusual research carried out by the small band of dedicated academics that has kept the quirky flag flying over the past century.

Each chapter reveals the secret psychology underlying a different aspect of our lives, from deception to decision-making, selfishness to superstition. Along the way, we will encounter some of my favourite pieces of strange but fascinating research. Experiments that have, for instance, involved stalling cars at traffic lights and measuring the amount of subsequent horn-honking; examined why there are a disproportionate number of marine biologists called Dr Fish; secretly analysed the type of people that take more than ten items through express lines in supermarkets; asked people to behead live rats with a kitchen knife; discovered whether suicide rates are related to the amount of country music played on national radio; and proved beyond all reasonable doubt that Friday 13th is bad for your health.

The majority of the research that you are about to

encounter has, until now, been hidden away in obscure academic journals. The work is serious science, and much of it has important implications for the way in which we live our lives, and structure our society. However, unlike the vast majority of psychological research, these studies have something quirky about them. Some use mainstream methods to investigate unusual topics. Others use unusual methods to investigate mainstream topics. All of them are the result of behavioural scientists misbehaving.

Let the quirkology begin.

1

What does your date of birth *really* say about you? The new science of chronopsychology

How the lives of mass murderers have been used to test astrology, whether you really are born lucky, why the rich and famous lie about the day they were born, and how some people are literally dying to save taxes.

Given that approximately one hundred million Americans read their daily horoscopes, and about six million have paid a professional astrologer to analyse their personality,[1] it is easy to argue that astrological beliefs have stood the test of time. Even world leaders are not immune from the lure of the soothsaying stargazers. Both Ronald and Nancy Reagan were fond of consulting with the cosmos, allowing astrologers to influence many aspects of their political lives, including the timing of international summits, presidential announcements, and the flight schedule of Air Force One.[2]

Over the years, a small number of highly dedicated scientists have investigated the relationship between people's lives and their date of birth. The work has involved studying mass murderers, trawling through millions of American tax returns, examining the birthdates of Premier League footballers, having over 20,000 people come online and assess their luck, and asking a 4-year-old child to predict the movement of international stock markets. Slowly but surely, the work has sifted fact from fiction to reveal the many ways in which our date of birth *really* influences the way we think and behave.

Prophets and profits

The British Association for the Advancement of Science (BAAS) was established in 1831 by the eminent Scottish scientist Sir David Brewster. The BAAS has several claims to fame. The term 'dinosaur' was first used at one of its meetings in 1841, and at their 1860 annual gathering, physicist Sir Oliver Lodge presented one of the first public demonstrations of wireless transmission. Also in 1860, they staged an infamous public debate about evolution between biologist T. H. Huxley and the Bishop of Oxford, Samuel Wilberforce (nicknamed 'Soapy Sam' because of his slipperiness during ecclesiastical debates). Rumour has it that during the debate Wilberforce turned to Huxley and asked: 'Is it on your grandfather's or grandmother's side that you claim descent from the apes?' Unfazed, Huxley turned to his colleagues and quietly muttered, 'The Lord hath delivered him into my hands,' before publicly declaring that he would rather be descended from an ape than a bishop.

Each year the BAAS coordinates a wonderful week-long national celebration of science, and in 2001 they invited me to conduct an experiment as part of the proceedings. After receiving the invitation, I happened to discover a newspaper article describing the latest fad in stargazing – financial astrology. According to the article, some soothsayers were claiming that the date of a company's formation could affect its future financial performance. If true, this had enormous implications for investors all over the world, and so I decided to discover whether heavenly activity really could predict the bottom line.

The experiment involved three participants – a financial astrologer, an experienced city analyst, and a young child. At the start of the test we gave them a notional £5,000 each, and asked them to invest the money as they thought best on the stock market. Then, over the course of a week, we tracked their choices. Who would make the wisest investments?

Finding astrologers to take part in these types of studies is notoriously difficult. The vast majority are unwilling to have their claims put to the test, and those that are interested rarely agree to the conditions associated with a scientific experiment. However, after a few dozen telephone calls, we found a professional financial astrologer who thought the project sounded fun, and was kind enough to accept the challenge.

Our remaining two guinea pigs proved easier to recruit. A quick Internet search and a couple of telephone calls brought to light an experienced city analyst who was also happy to throw his hat into the ring. Finally, a friend of a friend said that they would ask their daughter if she wanted to be our third and final participant. A bar of chocolate sealed the deal, and Tia, a 4-year-old girl from south-east London, with no investment experience, completed the team. When Barclays Stockbrokers, one of Britain's leading investment firms, agreed to adjudicate the contest, we were all set to go.

We allowed our three volunteers to invest their cash in any of the one hundred largest companies in the UK. Our financial astrologer carefully examined the formation date of the companies, and promptly plumped for a variety of different sectors, including communication- and technology-based stock (Vodafone, Emap, Baltimore Tech, and Pearson). Our investor drew on his seven years of extensive experience, and decided to invest mainly within the communications industry (Vodafone, Marconi, Cable & Wireless, and Prudential).

We wanted Tia's choices to be totally random, and she happily approved a cunning selection procedure involving a stepladder and a big pad of paper. At 11.55 a.m. on 15 March 2001, I found myself balancing precariously on the top of a six-foot stepladder in the marble foyer of Barclays Stockbrokers. Tia, and a small audience of Britain's top investors, were waiting patiently on the ground below. One of my hands gripped the ladder tightly, while the other held one hundred small pieces of paper, each bearing the name of a company. As the clock struck midday, I threw the papers high into the air and Tia randomly grabbed four of them as they gently fluttered to the ground. She carefully handed the four pieces to her mother, who announced that her daughter would invest in a high street bank (Bank of Scotland), a consortium of well-known drinks brands (Diageo), a financial services group (Old Mutual), and a leading supermarket chain (Sainsbury). The onlookers applauded, and Tia curtsied to her small but appreciative audience.

To be as fair as possible, we allowed our participants to change their investments a few days into our week-long experiment. Our financial astrologer again consulted the heavens, and swapped three of her choices, so that her final portfolio contained BOC, BAE Systems, Unilever, and Pearson. In one interview with journalists, she justified her decisions on the basis that these companies had a good planetary wind behind them.[3] Our expert investor chose to stick with his original selections. A second round of random paper-dropping left Tia with Amvescap, Bass, Bank of Scotland, and the Halifax.

At the end of the week, we regrouped at Barclays Stockbrokers and examined the results. It had proved an exceptionally turbulent week for the stock market, with billions of pounds wiped off the value of the world's leading companies. Strangely enough, neither of our experts had seen the crash coming. In line with this dramatic downward trend, all three of our participants had lost money. Bottom of the league came the financial astrologer, whose heavenly decisions resulted in a 10.1 per cent loss. The expert investor came a close second with a 7.1 per cent loss. Top of the class came Tia, with a loss of just 4.6 per cent.

Our investor didn't exactly display the kind of optimism commonly associated with city trading, telling journalists that he had confidently expected to finish last, and thought all along that Tia would win.[4] Our astrologer turned to the heavens to help explain her failure, noting that if she had known beforehand that Tia was a Cancerian she wouldn't have played against her.[5] Tia was remarkably modest about her win, saying that she couldn't explain her winning ways, and didn't even study science at nursery school.[6]

The *Sun* newspaper was rather taken with Tia's success and carried a full-page profile of her in their financial section, including her three top tips for those eager to play the markets: 'Money isn't everything – sweets are', 'Go to bed early', and 'Watch the growing market in kids' toys'.[7] *The Tonight Show with Jay Leno* expressed an interest in having

Tia on their programme, and I suspect that she was their only guest to decline on grounds of homework.

A week is not a long time in the world of finance, and so we decided to continue the experiment for a full year. It proved to be a difficult twelve months, with the global market showing an overall drop of 16 per cent. However, almost one year on from our original experiment, we asked Barclays Stockbrokers to reassess the value of the three portfolios. This time the differences were even more dramatic. Our investor had made a 46.2 per cent loss on his original investment. The financial astrologer did somewhat better, but still made a 6.2 per cent loss. Once again, Tia led the pack. In the face of a falling market she had managed to make a 5.8 per cent profit.[8]

I wasn't entirely surprised that our experts' predictions were less than impressive. This was not the first time that the wisdom of city analysts had come under scrutiny and been found wanting. In a similar Swedish study, a national newspaper gave $1,250 each to five experienced investors, and a chimpanzee named Ola. Ola made his choice by throwing darts at the names of companies listed on the Stockholm exchange. After a month, the newspaper compared the profits and losses made by each competitor. Ola had out-performed the financial wizards. Similarly, the *Wall Street Journal* regularly asks four investors to pick one stock apiece, and then randomly selects four other stocks by using Ola's dart-throwing technique. After six months, the paper compares the returns on the stocks selected by the experts with the 'dartboard portfolio'. The darts are often the more successful, and almost always beat at least one of the experts.

My test of financial astrology was not the first scientific examination of the alleged relationship between heavenly activity and earthly events. Similar work stretches back over decades, and has involved a series of unusual experiments, including work carried out by one of Britain's most prolific psychologists.

Heavenly predictions

Professor Hans Eysenck was arguably one of the most influential thinkers of the twentieth century, and, at the time of his death in 1997, was the living psychologist most frequently cited in scientific journals and magazines. Famous for liking the phrase, 'If it cannot be measured, then it does not exist', Eysenck spent much of his career trying to quantify aspects of the human psyche (including poetry, sexual behaviour, humour, and genius) that many believed to be beyond the grasp of science. He is, however, perhaps best known for his work on the analysis of human personality, and developed some of the most widely used personality questionnaires in modern-day psychology.

To fully appreciate Eysenck's astrological investigation, it is necessary to understand a little of his work into personality. Eysenck arranged for thousands of people to complete questionnaires about themselves, and then analysed the results using powerful statistical techniques designed to uncover the key dimensions on which people differed. The results revealed that people's personalities are not nearly as complex as they first appear. In fact, according to Eysenck, they vary on only a handful of fundamental dimensions, the two most important of which he labelled 'extroversion' and 'neuroticism'. The Eysenck Personality Inventory was designed to measure these traits, and contains about fifty statements. It asks people to indicate whether each statement describes them by circling either 'Yes' or 'No'.

The first of Eysenck's personality dimensions, extroversion, is all about the level of energy with which people approach life. High on the scale are the 'extroverts'. These people tend to be impulsive, optimistic, happy, enjoy the company of others, strive for instant gratification, have a wide circle of friends and acquaintances, and are more likely than others to cheat on their partner. At the other end of the scale come the 'introverts', who are far more considered, controlled, and reserved. Their social life revolves around a relatively small number of very close friends, and they prefer reading a good book to a night out on the town. Most people fall somewhere between these two extremes, and the Eysenck Personality Inventory measures people's level of extroversion–introversion by presenting them with statements such as 'I am the life of the party' and 'I feel comfortable around people'.

The second dimension, neuroticism, concerns the degree to which a person is emotionally stable. High scorers tend to be prone to worry, have low self-esteem, set themselves unrealistic targets or goals, and frequently experience feelings of hostility and envy. In contrast, low scorers are far more calm, relaxed and resilient in the face of failure. They are skilled at using humour to reduce anxiety, and sometimes even thrive on stress. The Eysenck Personality Inventory measures people's level of neuroticism using statements such as 'I worry about things' and 'I get stressed out easily'.

According to ancient astrological lore, six of the twelve signs of the zodiac are traditionally associated with extroversion (Aries, Gemini, Leo, Libra, Sagittarius, and Aquarius) and six with introversion (Taurus, Cancer, Virgo, Scorpio, Capricorn, and Pisces). Similarly, people born under the three 'earth' signs (Taurus, Virgo, and Capricorn) are seen as emotionally stable and practical, whilst those associated with the three 'water' signs (Cancer, Scorpio, and Pisces) should be far more neurotic.

To find out if this really were the case, Eysenck teamed up with a respected British astrologer named Jeff Mayo. A few years before, Mayo had founded the Mayo School of Astrology, and rapidly gained a large following of students across the globe. Just over 2,000 of Mayo's clients and students were asked to report their date of birth and complete the Eysenck Personality Inventory. Those sceptical about astrology expected the findings to reveal absolutely no relationship between participants' personalities and ancient astrological lore. In contrast, proponents of astrology were confident that the positions of the heavens at the time of birth would have a predicable impact on people's thinking and behaviour.

Much to the surprise of the sceptics, the results were perfectly in line with astrological lore. Those born under the signs traditionally associated with extroversion did have slightly higher extroversion scores than others, and those born under the three water signs obtained significantly higher neuroticism scores than those born under the earth signs.[9] The astrological journal *Phenomena* announced that these findings were 'possibly the most important development for astrology in this century'.[10]

But Eysenck became suspicious when he realized that the

participants in his study already had a strong belief in astrology. Most people who have such beliefs are well aware of the type of person that astrology predicts they are *meant* to be, and he wondered whether this knowledge had undermined the study. Could his results have been due to his participants thinking they had the type of personality that they knew was associated with their star sign? Could psychology, rather than the position of the planets at the time of their birth, account for his remarkable results?

Eysenck conducted two additional studies to explore this idea. The first involved people who were far less likely to have heard about the personality characteristics associated with different star signs – a group of 1,000 children. This time the results were dramatically different, and didn't match the patterns predicted by astrological lore: the children's levels of extroversion and neuroticism were completely unrelated to their star sign. To make absolutely certain, Eysenck ran a second birthdate–personality study with adults, but also assessed the amount that they knew about astrology. Those who knew a great deal about the effect that the planets should have on their personality did conform to the pattern predicted by astrology. In contrast, those who professed no knowledge showed no patterning. The conclusion was clear. The positions of the planets at the moment of a person's birth had no magical effect on personality. Instead, many people who were well aware of the personality traits associated with their sign, had developed into the person predicted by the astrologers.[11] When Eysenck presented these follow-up findings at a conference exploring science and astrology, his biographer noted that '. . . there was a strong feeling among some of the astrologers that Eysenck had first beguiled them with his patronage, and then betrayed them by bringing forward some ugly facts'.[12]

This is not the only time that researchers have discovered evidence of people becoming the person that they *ought* to be. In the 1950s, psychologist Gustav Jahoda studied the lives of the Ashanti people in central Ghana. According to tradition, every Ashanti child receives a spiritual name that is based upon the day they are born, and each day is associated with a set of personality traits. Those born on a Monday are referred to as *Kwadwo*, and are traditionally seen as quiet, retiring, and peaceful. Children born on a Wednesday are referred to as *Kwaku*, and are expected to be badly behaved. Jahoda was curious to discover whether this early labelling could have a long-term impact on the self-image, and lives, of the Ashanti children. To find out, he examined the frequency with which people born on different days of the week appeared in juvenile court records. The results showed that the label given to a child at birth affected their behaviour, with significantly fewer Kwadwos, and more Kwakus, appearing in the records.[13]

Did Eysenck's results cause millions to alter their belief in heavenly influence? Apparently not. Instead, many proponents of astrology argued that the star signs provided merely a very rough guide to a person's personality, and that real accuracy could be obtained only by carefully studying the precise moment that a person entered the world. It is a claim that has received a great deal of attention from researchers around the globe.

Time twins and Pogo the Clown

British researcher Geoffrey Dean is a quietly spoken, mild-mannered man who has dedicated his life to collecting, and collating any information that might allow him to assess the potential impact of the stars on human behaviour. He is in a unique position to carry out the work, being one of the very few scientific researchers in the world who used to earn his living as a professional astrologer.

In 2000, I was invited to speak at an international science conference in Australia, and was delighted to discover that Geoffrey was on the same bill. During his talk, Geoffrey described his latest and largest project: an investigation that he referred to as the 'definitive test' of astrology. Like so many good ideas, this one was very simple. According to the claims of astrologers, the position of the planets at a person's moment of birth predicts their personality, and key events in their lives. If this is the case, then people born at the same moment, and in the same place, should be almost identical to one another. In fact, they should, as Geoffrey noted, be 'time twins'.

There is some anecdotal evidence to support the idea. In the 1970s, astrological researchers trawled through a database of births, and noted that some people born within a few days of one another lead surprisingly similar lives. For instance, the French champion bicycle racers Paul Chacque and Leon Lével were born on 14 July 1910 and 12 July 1910 respectively. They were both highly successful in 1936, with Chacque winning the Bordeaux–Paris section of the Tour de France, and Lével winning the two mountain sections of the same race. In March 1949, Lével died when he fractured his skull in an accident on the Parc des Princes track. Chacque died from a similar injury, on the same track, in September of the same year.[14]

Intriguing though cases like this are, they could simply be the result of chance, and so Geoffrey decided to carry out more systematic work into the alleged phenomenon. He managed to uncover a database containing the details of just over 2,000 people born in London between 3 and 9 March in 1958. The database had been put together by a group of researchers studying people as they progressed through life, and contained the results of intelligence tests and personality questionnaires, administered at the ages of 11, 16 and 23. The precise time of birth for each person had been carefully recorded, with over 70 per cent of them being born within five minutes of one another. Geoffrey arranged the group in order of birth, and moved down the list, calculating the degree of similarity between each pair of people. Once again, the sceptics and proponents made very different predictions about what he would find. The sceptics thought that there would be no relationship between the test results of each pair of people on the list. In contrast, the astrologers expected to see the type of striking similarities found between the personalities of identical twins.

This time, the sceptics were right. Geoffrey found little evidence of similarity between his time twins. People born at five minutes past eleven on 4 March 1958 were no more similar to their time twin born moments later than another person born days later.[15]

Geoffrey has carried out many tests like this and the results have one thing in common – none of them provide any support for the claims of astrology.[16] As a result, he sometimes describes himself as 'the most hated person in astrology', and is seen as something of a turncoat by modern-day astrologers – a man who has gone to the dark side by publicly declaring his scepticism about the impact of the heavens on our lives.

Geoffrey's research tends to be methodologically similar to the work carried out by Hans Eysenck, in that it usually involves the examination of large amounts of data in search of the type of patterns predicted by astrology. This is not, however, the only approach to testing the accuracy of heavenly predictions. Other researchers have examined the claims made by individual astrologers. One of the most unusual and striking examples of this approach was reported by a group of American researchers in the late 1980s, in a provocatively entitled article, 'Astrology on Death Row'.[17]

The researchers first found out the birth time, date, and place of the notorious serial murderer John Gacy. Gacy was a sadistic killer who received twelve death sentences and twenty-one life terms for the torture and killing of thirty-three men and boys. By dressing up as Pogo the Clown, and performing at children's birthday parties in his spare time, Gacy may have given rise to the notion of the 'evil' clown. One of the researchers visited five professional astrologers, and presented Gacy's details as his own. The researcher explained to each astrologer that he was interested in working with young people, and asked for a general personality reading and some career advice. The astrologers got it badly wrong. One encouraged the researcher to work with young people because he could 'bring out their best qualities'. Another analysed the information provided and confidently predicted that the researcher's life would be 'very, very positive'. A third said that he was 'kind, gentle, and considerate of others' needs'.

The work of Hans Eysenck, Geoffrey Dean, and others show that heavenly predictions often fall far short of the mark. In doing so, they leave us with a bigger mystery: Why do so many people believe in astrology?

Professor Bertram Forer and the nightclub graphologist

In the late 1940s, Professor Bertram Forer was busy devising novel ways of measuring personality. One evening Forer visited a nightclub, and was approached by a graphologist who offered to determine his personality on the basis of his handwriting. Forer declined the offer, but the chance encounter made him want to discover why large numbers of people were impressed with astrologers and graphologists. Forer could have carried on with his normal academic research. Curiosity got the better of him, however, and he decided to carry out an unusual experiment. It was an experiment that was to make him famous long after his mainstream work on personality faded into obscurity.

Forer had the students in his introductory psychology class complete a personality test.[18] One week later, each student was handed a sheet of paper and told that it contained a short description of their personality based on their test scores. Forer asked the students to examine the description carefully, assign it an accuracy rating by circling a number between 0 (poor) and 5 (perfect) on the sheet of paper, and then raise their hand if they thought the test had done a good job of measuring their personality.

Let's turn back the hands of time, and restage the experiment. Here is one of the descriptions that would have been handed to students in Forer's study. Read it through and see if you think it is a fairly accurate description of your personality:

> You have a need for other people to like and admire you, and yet you tend to be critical of yourself. While you have some personality weaknesses you are generally able to compensate for them. You have considerable unused capacity that you have not turned to your advantage. Disciplined and self-controlled on the outside, you tend to be worrisome and insecure on the inside. At times you have serious doubts as to whether you have made the right decision or done the right thing. You prefer a certain amount of change and variety and become dissatisfied when hemmed in by restrictions and limitations. You also pride yourself as an independent thinker, and do not accept others' statements without satisfactory proof. But you have found it unwise to be too frank in revealing yourself to others. At times you are extroverted, affable, and sociable, while at other times you are introverted, wary, and reserved. Some of your aspirations tend to be rather unrealistic.

Forer's students read the description, made their rating, and, one by one, started to raise their hands into the air. After a few moments, he was surprised to see virtually all of the students with their hands up. Why was Forer so amazed?

As is sometimes the case with psychology experiments, Forer had not been entirely honest with his guinea pigs. The personality description that he had handed them was not based on their test scores. Instead, it came from a news-stand astrology book that he had picked up a few days before. More importantly, *every student had received exactly the same personality description* – the description that you read a few moments ago.

Forer had simply gone through the astrology book, selected about ten or so sentences from different astrological readings, and glued them together to make a single description. Despite all being given the same personality description, 87 per cent of students had circled either 4 or 5 on the rating scale, indicating that they were extremely impressed with the accuracy of what they had read. The reading Forer created has become world famous, and has been used in thousands of psychology experiments and television shows.

Forer's results solved the mystery that had been bugging him since his chance meeting with the graphologist. Astrology and graphology do not actually need to *be* accurate in order to be *seen* as accurate. Instead, all you have to do is give people a very general statement about their personality, and their brains will trick them into believing that it is insightful.

Immediately after conducting his study, Forer told his students that they had all received the same personality description, explained that the exercise had been 'an object lesson to demonstrate the tendency to be overly impressed by vague statements', and pointed out the 'similarities between the demonstration and the activities of charlatans'. Apparently, most of Forer's students were not too upset about being exposed as a tad gullible. Many of them bestowed on the psychology experiment the greatest honour that a student can give, asking Forer for a copy of the personality description so that they could play the same trick on their friends.

Most psychologists would have left it there, but Forer devised one last twist in a final attempt to inflict further humiliation on his long-suffering class.

Forer wondered whether his students would want to see themselves as astute, streetwise, and smart. If so, would their acceptance of vague personality statements have presented a real challenge to this aspect of their self-identity? Moreover, rather than go through the painful process of seeing themselves as they really are, would many take the easy option of simply denying to themselves that they were taken in by the demonstration?

Three weeks later, Forer told his class that he had inadvertently erased their names from the rating sheets, and asked them to jot down honestly the ratings that they had assigned the original description. In reality, he had not lost the names at all, and so was able to compare the ratings each student had originally given the description with the rating they subsequently claimed that they had given it. Half of the students who had *originally* indicated that they thought the description was 'perfect' (assigning it the maximum score of 5) subsequently claimed that this was not the case and said that they had given it a lower rating. It seems that the most gullible people would rather fool themselves than face up to their gullibility.

Enter Phineas Taylor Barnum

In the 1950s, psychologist Paul Meehl christened Forer's original finding 'the Barnum Effect', after American showman Phineas Taylor Barnum, who once famously said that any good circus should have something for everyone.[19] Years of research have shown that almost everyone is prone to the Barnum Effect – men and women, young and old, believers in astrology and sceptics, students, and even personnel managers.[20]

One of the most interesting follow-up studies was conducted by French researcher Michel Gauquelin.[21] Gauquelin sent the birth details of Dr Marcel Petiot, a notorious French mass murderer, to a firm that used high-tech computers to generate allegedly accurate horoscopes. During the Second World War, Petiot told his victims that he was able to help them escape from Nazi-occupied France, but instead administered a lethal injection and watched them slowly die. Petiot later pleaded guilty to nineteen murders, and was guillotined in 1946. The computerized horoscope managed to miss all of the rather grisly aspects of Petiot's life, and instead generated the same type of bland Barnum statements that had been used to such great effect by Forer, including:

> His adaptable and pliant character expresses itself through skill and efficiency; his dynamism finds support in a tendency towards order, control, balance. He is an organized and organizing person socially, materially and intellectually. He may appear as someone who submits to social norms, fond of propriety and endowed with a moral sense which is comforting – that of a worthy, right-thinking, middle-class citizen.

Although Petiot was executed in 1946, the horoscope predicted that between 1970 and 1972 he would experience 'a tendency to make commitments regarding his romantic life'.

Inspired, Gauquelin then placed an advertisement in a well-known newspaper, offering free, computer-generated horoscopes. Over 150 people from all over France responded, and Gauquelin sent each one the reading based on the birth details of Petiot. He also asked them to rate the degree to which the horoscope presented an accurate description of their personality. Ninety-four per cent of recipients said that it was accurate. One person wrote to Gauquelin, noting, 'The work done by this machine is marvellous . . . I would go so far as to say extraordinary,' whilst another wrote, 'It is absolutely bewildering that an electronic machine is able to probe people's character and future.' Some people were so impressed that they offered to pay Gauquelin for a more detailed analysis.

So why are so many people taken in by these types of readings?

People endorse many of the statements because they are true for the vast majority of the population. After all, who hasn't had serious doubts about an important decision, would deny wanting other people to admire them, or doesn't strive for a sense of security? Even specific-sounding statements can be true of a surprisingly large percentage of the population. A few years ago by a colleague of mine, psychologist Susan Blackmore, surveyed just over 6,000 people, asking them about the sorts of seemingly specific statements that crop up in psychic readings, such as, 'You have someone in your family named Jack'.[22] She discovered that about a third of people have a scar on their left knee, another third own a tape or CD of Handel's *Water Music*, a fifth had a 'Jack' in the family, and about one in ten had spent the previous night dreaming about someone they hadn't seen for years. It seems that many Barnum statements appear accurate because most people tend to think and behave in surprisingly predictable ways.

Then there is the 'flattery effect'. Most people are more than willing to believe anything that puts them in a positive light, and thus endorse statements suggesting that they have a great deal of unused capacity or are independent thinkers. This effect explains why half of the population is especially accepting of astrology. The twelve signs of the zodiac are traditionally split into the six 'positive' signs (Aries, Gemini, Leo, Libra, Sagittarius, and Aquarius) and six 'negative' signs (Taurus, Cancer, Virgo, Scorpio, Capricorn, and Pisces). The traits associated with the positive signs tend to be more favourable than those associated with the negative signs. Those born into Libra are traditionally seen as the type of people who seek peace and beauty, whilst Taureans are viewed as more materialistic and easily upset. Psychologist Margaret Hamilton from the University of Wisconsin asked people to give their date of birth, and rate the degree to which they believed in astrology on a seven-point scale. As predicted by the 'flattery effect', those born under 'positive' signs were significantly more likely to believe in astrology than those born under 'negative' signs.[23]

The work of Forer, and those who have followed in his footsteps, demonstrates how horoscopes have fooled millions of people over thousands of years. Astrologers can produce any old tosh and, providing it is sufficiently vague and flattering, the majority of people will tick the 'highly accurate' box. So, given that the scientific evidence in favour of astrology is less than overwhelming, it would be tempting to conclude that there is no real science associated with a person's date of birth.

Tempting, but wrong.

The scientific study of time and mind

Chronopsychology is a new, and still relatively obscure, scientific discipline devoted to the study of time and mind. Much of the work in this area is concerned with circadian rhythms, shift work, and jet lag.

In 1962, French caver and geologist Michel Siffre decided to spend two months entirely below ground, tracking the movement of a glacier through an underground ice cave.[24] Rather than simply sit there taking measurements and twiddling his thumbs, Siffre made the most of his subterranean isolation by carrying out a unique experiment into the psychology of time. Siffre decided not to take any type of clock with him into the cave, and so forced himself to rely entirely on his own body clock to decide when to fall asleep and when to be awake. Siffre's only link to the outside world was a telephone that provided a direct line to a team of researchers above ground. Siffre called the team whenever he was going to sleep and when he woke up, and from time to time during his waking hours. Each time, the experimenters above ground did not give him any indication of the real time. Deprived of any daylight for over sixty days in his small nylon tent 375 feet below ground, Siffre's telephone calls showed that his ability to judge time became radically distorted. Towards the end of the experiment he would telephone the surface, convinced that only an hour had gone by since his previous call – whereas in reality, several hours had elapsed. When he was brought back out of the cave after two months, Siffre was convinced that the experiment had been terminated early, and that it was only in its thirty-fourth day. The experiment provided a striking illustration of how daylight helps our internal clocks to keep accurate time.

Other chronopsychological work has examined ways of minimizing the effects of perhaps the most common and annoying form of disruption faced by the modern-day body clock – jet lag. One of the more unusual and controversial studies in this area was carried out in the late 1990s by Scott Campbell and Patricia Murphy of Cornell University, and involved shining lights on the back of people's knees.[25] Previous work had shown that shining light into people's eyes fools the brain into speeding up or slowing down their biological clock, and so can help to overcome the effects of jet lag. Campbell and Murphy wondered whether people might detect similar signals from other parts of their body. As the back of the knee contains a large number of blood vessels very close to the surface of the skin, they decided to test their hypothesis by applying light to the region using specially designed halogen lamps. In a small-scale study, they found evidence that light applied to the back of the knee matched the biological-clock-altering ability of light shone directly into the eyes.

So what is the link between the ideas underpinning astrology and this fascinating scientific area of study? Not all chronopsychology is about spending months in caves and shining light on the backs of people's knees. Another branch of this obscure discipline has involved a small number of scientists examining the subtle influence that people's birthdays may exert over the way in which they think and behave.

The concept behind this unusual branch of behavioural science is beautifully illustrated by the work of Dutch psychologist Ad Dudink.[26] After analysing the birthdays of just under 3,000 English professional football players, Dudink found that twice as many were born between September and November as were born between June and August. It seemed that a person's date of birth predicted their sporting success. Some may have viewed the result as compelling evidence for astrology, arguing that the planetary positions associated with Virgo, Libra, Scorpio, and Sagittarius play a key role in creating successful athletes. There is, however, a more interesting and down-to-earth explanation for Dudink's curious finding.

At the time of his analysis in the early 1990s, budding English footballers were eligible to play professionally only if they were at least 17 years old when the season started, which was in August. Potential players born between September and November would therefore have been about ten months older, and more physically mature, than those born between June and August. These extra few months proved a real bonus when it came to the strength, endurance and speed needed to play football, with the result that those born between September and November were more likely to be picked to play at a professional level.

Years of research have revealed an overwhelming amount of evidence for the effect in many different sporting arenas. Regardless of when a sporting season starts, there is an excess of athletes whose month of birth falls in the first few months of that season. From American Major League baseball to

British county cricket, Canadian ice hockey to Brazilian soccer, the month of birth of athletes is related to their sporting success.[27]

Such chronopsychological effects are not limited to the lives of professional athletes. They also influence a factor that plays a key role in everyone's life – their luck.

Born lucky?

Are you lucky or unlucky? Why do some people always seem to be in the right place at the right time, whilst others attract little but bad fortune? Can people change their luck? About ten years ago, I decided to answer these types of intriguing questions by carrying out research into the psychology of luck. As a result, I have worked with about 1,000 exceptionally lucky and unlucky people drawn from all walks of life.[28]

The differences between the lives of the lucky and unlucky people are as consistent as they are remarkable. Lucky people always seem to be in the right place at the right time, fall on their feet, and appear to have an uncanny ability to live a charmed life. Unlucky people are the exact opposite. Their lives tend to be a catalogue of failure and despair, and they are convinced that their misfortune is not of their own making. One of the unluckiest people in the study is Susan, a 34-year-old care assistant from Blackpool. Susan is exceptionally unlucky in love. She once arranged to meet a man on a blind date, but her potential beau had a motorcycle accident on the way to their meeting, and broke both of his legs. Her next date walked into a glass door and broke his nose. A few years later, when she had found someone to marry, the church in which she intended to hold the wedding was burnt down by arsonists just before her big day. Susan has also experienced an amazing catalogue of accidents. In one especially bad run of luck, she reported having eight car accidents in a single fifty-mile journey.

I wondered if good and bad fortune really was chance, or if psychology could account for these dramatically different lives, and so designed a series of studies to investigate. In one especially memorable study, I gave some volunteers a newspaper and asked them to look through it and tell me how many photographs were inside. What I didn't tell them was that halfway through the newspaper I had placed an unexpected opportunity. This 'opportunity' took up half of the page and announced, in huge type, 'Win £100 by telling the experimenter you have seen this.' The unlucky people tended to be so focused on the counting of the photographs that they failed to notice the opportunity. In contrast, the lucky people were more relaxed, saw the bigger picture, and so spotted a chance to win £100. It was a simple demonstration of how lucky people can create their good fortune by being more able to make the most of an unexpected opportunity.

Results like this revealed that the volunteers were making much of their good and bad luck by the way they were thinking and behaving. The lucky people were optimistic, energetic, and open to new opportunities and experiences. In contrast, the unlucky people were more withdrawn, clumsy, anxious about life, and unwilling to make the most of the opportunities that came their way.

Some of my most recent work in the area has taken a chronopsychological turn, exploring whether there is any truth to the old adage that some people are actually born lucky.[29] The project had its roots in a rather curious email that I received in 2004 from Professor Jayanti Chotai, a researcher at the University Hospital in Umeå in Sweden.

Much of Jayanti's work examines the relationship between people's date of birth and many different aspects of their psychological and physical well-being. In one of his studies, Jayanti had asked a group of about 2,000 people to complete a questionnaire measuring the degree to which they described themselves as sensation-seekers, and then looked to see whether there was any relationship between their questionnaire scores and month of birth.[30] Novelty and sensation-seeking are fundamental aspects of our personalities. Very high sensation-seekers can't stand watching movies that they have seen before, enjoy being around unpredictable people, and are attracted to dangerous sports such as mountain-climbing or bungee rope-jumping. In contrast, very low sensation-seekers like to see the same movies again and again, enjoy the comfortable familiarity of old friends, and don't like visiting places that they haven't been to before. Jayanti's results suggested that sensation-seekers tended to be born in the summer, whilst those more comfortable with the familiar were likely to be born in the winter.

Jayanti's email explained that he had read about my work on the link between personality and luck, and wondered whether some people were actually born lucky. It was an intriguing idea, and the two of us decided to team up and discover if there was any truth in the expression.

Jayanti's previous work had suggested that the relationship between people's month of birth and personality was real, but small. To detect very small effects we needed to carry out an experiment involving thousands of people. We also knew that that wasn't going to be easy. Getting even a few hundred students to participate in research is often problematic, and we needed thousands of people from all walks of life to take part if we were to have any chance of finding what we were looking for. Luckily, help was at hand.

The Edinburgh International Science Festival in Scotland is one of the oldest science festivals in the world, and the largest in Europe. By making the experiment part of the festival, it would stand a good chance of attracting the large numbers of people needed. The festival organizers gave us the green light, and we set up a simple website where people could enter their date of birth, and then answer a standard questionnaire that I had devised to assess their level of luck.

Carrying out large-scale public experiments is always an unpredictable affair. Unlike laboratory research, you only have one opportunity to get it right, and you never know whether people will give up their time to participate. However, our study attracted the public's imagination, and quickly spread around the globe. Within a few hours of going live we had hundreds of people using the site. By the end of the festival we had collected data from over 40,000 people.

The results were remarkable. Jayanti had found those with summer births to be the risk-takers. Our results revealed that those born in the summer (March to August) also rated themselves as luckier than those born in the winter (September to February). The resulting levels of luckiness across the twelve months showed an undulating pattern that peaked in May and was at its lowest in October (see graph opposite). Only June failed to fit the pattern – something that we put down to a statistical blip.

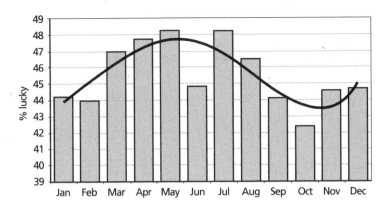

Born lucky results: the percentage of lucky people born in each month of the year

There are lots of possible explanations for an effect like this. Many of them revolve around the notion that the ambient temperature is colder in the winter than in summer. Perhaps because babies born in winter enter a harsher environment than those born in the summer, they remain closer to their care-givers, and so are less adventurous and lucky in life. Or perhaps women who give birth in late winter have had access to different foods than those who give birth in the summer, and this affects the personality of their offspring. Whatever the explanation, the effect is theoretically fascinating, and suggests that the temperature around the time of birth has a long-term effect on the development of personality.

But before accepting any of the temperature-related explanations, it is important to rule out other possible mechanisms. Maybe the effect is nothing to do with temperature, but instead concerns some other factor that varies across the year. Proponents of astrology might argue that heavenly activity was affecting people's personalities, and that the alignment of the planets and stars during the summer months somehow gives rise to a luckier life.

The only way of assessing the competing explanations would be to repeat the study in a place where there is a different relationship between temperature and the months of the year. If the temperature-related explanations were correct, then the warmer months should still be associated with lucky births. If the astrological explanations were right, then May, June, and July, etc., should emerge as the lucky months.

People living in the earth's southern and northern hemispheres experience almost exactly the opposite relationship between temperature and month of the year. In the northern hemisphere, it is hot in June and cold in December. In the southern hemisphere this pattern is reversed, and so June is a winter month whereas December is gloriously hot. Because of this, I decided to compare the temperature-related and astrological interpretations of the 'born lucky' effect by restaging the experiment on the other side of the earth.

The city of Dunedin is situated on the south-eastern coast of New Zealand's South Island, and is home to a biennial science festival. In 2006, I received an email from the organizers of the New Zealand festival. They had heard about the need to repeat the 'born lucky' experiment in the southern hemisphere, and wondered whether I would like to carry out the study again at their festival. I was soon heading for New Zealand.

The national media in New Zealand and Australia reported the Born Lucky 2 experiment, and helped attract people to another specially designed website. Within a few days, over 2,000 people provided their month of birth, and rated the degree of luck in their lives. The results supported a temperature-related explanation for the effect. Those born in the summer months of the southern hemisphere (September to February) considered themselves significantly luckier than those born in winter (March to August). Again, an undulating pattern emerged across the twelve months, but this time it peaked in December and was at its lowest in April.

Experiments like the Born Lucky studies suggest that the month in which people are born exerts a small, but real, influence over the way in which they behave. But other research has investigated exactly the opposite effect, namely how people's behaviour influences the alleged birthdate of themselves and others.

The chronopsychology of tax evasion and lying clergy

The American tax system is set up in such a way that a family whose child is born on 31 December receives tax benefits for the previous twelve months, whereas a child born on 1 January doesn't. Because of this, there is a considerable financial incentive for parents expecting a child around the end of the year to ensure that the baby is born before midnight on 31 December. Although parents are unable to predict accurately the exact date of a natural birth, they are able to manipulate the child's date of birth by asking for an induced labour or a Caesarean section.

Would parents really manipulate their child's date of birth simply to obtain tax benefits? To find out, Stacy Dickert-Conlin from Syracuse University and Amitabh Chandra from the University of Kentucky analysed the American natality records between 1979 and 1993.[31] Focusing on the final seven days of December and the first seven of January, they found a sudden and significant peak in the number of births at the end of December in all but one of the years.

The professors then dug deeper into the data to find out whether this unusual pattern really was caused by parents trying to increase their benefits. They carefully analysed the individual family circumstances of nearly 200 births that took place in either the week before, or the week after, 1 January. For each birth they came up with two figures – the tax benefits associated with a December birth and the benefits associated with a January birth. The results revealed that the families of babies born in the last week of December had significantly more to gain financially from a December birth than the families of babies born in the first week of January. This was the clincher: compelling statistical evidence that parents were manipulating the date of birth of their offspring for financial gain.

There is, of course, a much easier way of manipulating your date of birth. A way that escapes the need for an induced labour, or a Caesarean section. Lie.

American actress Lucille Ball once famously said that the secret of staying young is to 'live honestly, eat slowly, and lie about your age'. She should know. Ball's actual date of birth was 6 August 1911, but through much of her career she claimed to have been born in 1914. Ball is far from being the only Hollywood legend to have lied about their age. Nancy Reagan claimed to be two years younger than she actually was, and even published an incorrect date of birth in her autobiography. Hollywood comedian Gracie Allen was so secretive about her age that even her husband, fellow performer George Burns, didn't know her actual date of birth. Various sources claim that Allen was born on 26 July in 1894, 1895, 1897, 1902, or 1906. Throughout her life, Allen claimed that her birth certificate was destroyed in the 1906 San Francisco earthquake, despite the earthquake occurring a few months *before* her alleged birthday. When asked about the discrepancy, Allen allegedly remarked, 'Well, it was an awfully big earthquake.'

It is not difficult to figure out the psychology behind such minor manipulations. In a society that places great value on the beauty of youth, it perhaps isn't surprising that many people wish to appear younger than they actually are. But would eminent and allegedly upstanding members of society be equally devious about even the *day* that they were born?

To find out, Professor Albert Harrison from the University of California, Davis, and his colleagues carefully worked their way through more than 9,000 biographical entries in several different volumes of *Who's Who* and *Who Was Who*, counting the number of people born either directly on, or three days either side of, the best-known dates in the American calendar: Independence Day (4 July), Christmas Day (25 December), and New Year's Day (1 January).[32] By chance alone, there should be roughly the same percentage of eminent people born on an auspicious day as on any one of the three days either side of that date. However, something strange was going on. Statistically, more people were born on Independence Day, Christmas Day, or New Year's Day than on one of the three days either side of these dates. The likelihood of this distribution by chance is hundreds to one, suggesting that some eminent individuals were misreporting their date of birth to a biographer in order to associate themselves with a day of national importance.

Harrison believes that the effect is due to an unusual psychological phenomenon called 'Basking In Reflected Glory' or, as many researchers refer to it, the 'BIRG' effect.

BIRG is a commonplace occurrence. We often hear people proudly announcing that they went to the same school as a well-known celebrity, or were one of the first to see a film that has just won an Oscar ('Guess who I had in the back of my cab yesterday?'). It even affects our everyday language. Psychologists secretly studying conversations on college campuses noticed big differences in a student's comments when their football team either won or lost a match. People were keen to associate themselves with their team's victory ('*we* won'), and just as eager to distance themselves from defeat ('*they* lost'). Harrison believes that the rich and famous were prepared to misreport their birthday to bask in the reflected glory of well-known days of the year. This interpretation of the data is supported by anecdotal evidence. The world-famous jazz musician Louis Armstrong claimed to have been born on 4 July. However, music historian Tad Jones has examined Armstrong's birth records and discovered that he was actually born on 4 August. Professor Harrison's results suggest that this is far from the only case of a celebrity blowing his own trumpet.

To further investigate the BIRG effect within *Who's Who* and *Who Was Who*, Harrison and his team focused their attention on the occupation that was most clearly associated with one of the dates – the clergy and Christmas. Working back through their data, they classified each member of the clergy into one of two groups. 'Eminent clergy' were those who listed their occupation as bishop or above, whilst 'non-eminent clergy' consisted of everyone else. By chance alone,

one would expect roughly the same percentage of both groups to be born on Christmas Day. In fact, significantly more eminent than non-eminent clergy claimed to share a birthday with Christ, perhaps supporting the idea that the higher you go in the clergy, the more you feel the need to move closer to Jesus.

Perhaps we are being a little harsh on the eminent folk involved in Harrison's analysis. In much the same way as some parents change the date of their offspring's birth to save a few dollars, others may have been so eager to see their children excel in life that they deliberately misreported the time of their birth to make the event seem more auspicious. Modern-day hospital births make such misreporting problematic, but in days gone by, parents reported their children's births verbally to local registry offices, making deception much easier. The mother of eminent mystery writer Georges Simenon confessed to this type of manipulation, reporting her son's birthday as being a day earlier than Friday 13 February 1903, because she considered the thirteenth to be 'too hard a fate for her sweet newborn baby'. If this interpretation of the results is valid, then it would be wrong to conclude that high-ranking clergy are more likely to lie than low-ranking clergy. Instead, the evidence would suggest that it is the *parents* of high-ranking clergy who are especially deceptive. Perhaps this represents one of the very few instances in which there is empirical evidence to support the Biblical notion that the sins of the fathers are visited upon their sons.

Some researchers now believe that parental lying may help account for a mystery that has baffled scientists for decades: a mystery that has come to be known as the 'Mars Effect'.

The Mars Effect

In addition to sending out horoscopes based around the birth-date of a mass murderer to innocent members of the public, French researcher Michel Gauquelin tested many aspects of astrology. According to astrological lore, being born when certain planets are high in the sky is a good omen, and suggests that the individual concerned will be eminent in his or her chosen career. In the 1950s, Gauquelin tested this notion by plotting the star charts of 16,000 people listed in a leading nineteenth-century French biographical dictionary. To his amazement, he discovered that certain planets were more likely to be above the horizon at the time of their births. For over fifty years this evidence, which has come to be known as the Mars Effect, puzzled even the most sceptical of thinkers. One researcher remarked that, 'It is probably not putting it too strongly to say that everything hangs on it,' and Hans Eysenck noted that if the 'results are ever shown to be spurious then, relatively speaking, the positive evidence that remains for astrology is weak'.[33] Then, in 2002, Geoffrey Dean, the researcher who carried out the 'time twins' experiment, undertook a remarkable piece of scientific detective work.[34]

During the nineteenth century, many in the French upper classes held a strong belief in astrology, and had ready access to popular almanacs that showed the exact position of the planets throughout each day. In addition, parents reported the time and date of their children's births verbally to local registry offices, rather than the data being officially and accurately recorded by doctors and midwives. Dean uncovered evidence to suggest that some parents were so eager to see their children excel in life that they deliberately misreported the date of their birth to make the event seem astrologically 'auspicious'. Such parents could then subsequently provide their children with the schooling and other resources required to turn these alleged 'heavenly predictions' into self-fulfilling prophecies. In short, Dean's work suggests that the Mars Effect may have little to do with astrology, and much more to do with a quirky piece of social history.

So far, we have been exploring how, and why, people manipulate their birthdate. An even stranger aspect of chronopsychology, however, has examined a rather more morbid topic – the relationship between the date of birth and the time of death.

Chronopsychology and the grim reaper

Sociologist David Phillips at the University of California, San Diego, is a man fascinated by death. Unlike many medical researchers, who are concerned with why people die, Phillips is more concerned with when. Specifically, he is interested in whether people are able to postpone their death until after a moment of important emotional significance. He has devoted his entire career to the topic, starting in 1970, when he published a doctoral dissertation with the curious title 'Dying as a form of social behaviour'.

Phillips is intrigued by the notion that people can exert enough control over their bodies to delay their demise for a small, but vital, period of time. Just long enough, it seems, to allow them to enjoy an important national or personal event.

There is certainly some anecdotal evidence to support the notion. Charles Schulz, the multi-millionaire cartoonist and creator of the 'Peanuts' strip, died on the eve of the publication of his last comic strip. The final cartoon contained a farewell letter signed by Schulz. Also, no less than three American presidents, John Adams, Thomas Jefferson, and James Monroe, all died on 4 July, thus raising the intriguing possibility that they held on long enough to ensure an auspicious date of death.

In one piece of research, Phillips examined whether people are more likely to die directly after a national holiday. There seemed little point in looking at mortality rates immediately before and after Christmas, as any significant rise in reported deaths could have been caused by the fact that the temperature tends to decrease throughout December. Rather than try to convince entire nations to celebrate Christmas in randomly determined months, Phillips searched for another national festival that took place at a different time each year. That is when he found the Chinese Harvest Moon Festival. This highly traditional celebration involves the senior woman of the household directing her daughters in the preparation of an elaborate meal, and moves around the calendar from year to year. An examination of Chinese death records around the event showed that the death rate dipped by 35 per cent in the week before the festival, and peaked by the same amount in the week after.[35]

One of Phillips' largest studies investigated whether people's date of birth influenced their date of death.[36] Analysing almost three million Californian death certificates between 1969 and 1990, Phillips reported that women are more likely to die in the week following their birthday than any other week of the year. In contrast, men are more likely to pass away in the week *before* their birthday. Phillips argued that this may be due to women looking forward to their birthday as a time of celebration, whereas men are more likely to use the coming occasion to take stock of their lives, realize how little they have achieved in life, become stressed, and therefore increase their chances of dying. According to Phillips, these findings are not due to seasonal fluctuations, people misreporting the information on their death certificates, putting off life-threatening surgery, or committing suicide. Instead, he argues that the data supports the notion that some people are capable of 'willing' themselves to live longer, or cutting short their lives.

The idea is highly controversial, and has attracted a great deal of debate.[37] Some researchers have been able to replicate the types of patterns found by Phillips and his team, whilst others have failed to find such results or attacked the methods used to conduct the work. Nevertheless, the idea that psychological factors influence physical well-being is supported by work showing a relationship between people's optimism and their health. For instance, in 1996, a team of researchers investigated the link between healthy thinking and longevity among 2,000 Finnish men. The team classified participants into three groups – a 'pessimistic' group who expected the future to be bleak, an 'optimistic' group who had much higher expectations about the future, and a 'neutral' group whose expectations were neither especially positive nor negative. They then monitored the groups over a six-year period, and found that the men in the 'pessimistic' group were far more likely to die from cancer, cardiovascular disease, and accidents than those in the 'neutral' group. In contrast, those in the 'optimistic' group exhibited a far lower mortality rate than those in the other two groups.[38]

Phillips is not the only researcher to investigate the strange factors that may influence the precise moment people meet the grim reaper. A paper published in the *Review of Economics and Statistics* in 2003 explored whether people's tax liability influenced their date of death,[39] combining Phillips' ground-breaking approach to death with the possibility (discussed earlier in the chapter) that parents may manipulate their off-spring's date of birth to obtain tax incentives. In this paper, Wojciech Kopczuk from the University of British Columbia and Joel Slemrod from the University of Michigan wondered whether people might die at a moment that is financially most beneficial to those left behind.

To discover if this was the case, they analysed the pattern in reported deaths around the time of significant changes to the American tax system. There were thirteen major changes to the tax laws between introduction of the tax system in 1916 and the present day, with eight of them resulting in increases in the rate of tax and five resulting in decreases. There was a period of approximately a week between the changes being announced in the media, and coming into effect. Analysing the reported number of deaths in the two weeks either side of each change, the researchers found evidence of an *increase* in the death rate just *before* a rise in the tax rate came into effect, and a *decrease* in the death rate just *after* a drop in the tax rate. This suggests, as indicated in the title of their paper, that some people may indeed be 'Dying To Save Taxes'.

This is not, however, the only interpretation of their data. Deaths are often reported by relatives who are likely to inherit the deceased's estate, and thus have a vested interest in reducing their tax liability. Consequently, the effect may be evidence of people misreporting the day that their wealthy loved one actually died, or, in the worst-case scenario, murder.

2

Trust everyone, but always cut the cards: The psychology of lying and deception

Exploring the language of lying with Hollywood star Leslie Nielsen, the link between freshly guillotined heads and the human smile, Ronald Reagan and the story that never was, the results of the mysterious Q-test, and the dark side of human suggestibility.

When I was 8 years old, I saw something that changed my life.

My grandfather handed me a marker pen, and asked me to write my initials on a coin. He carefully placed the coin on his palm, and closed his hand. After gently blowing on his fingers, he opened his hand, and the coin had mysteriously vanished. Next, he reached into his pocket and took out a small tin box that was sealed with several elastic bands. My grandfather handed me this rather strange-looking package, and asked me to remove the elastic bands and open the box. The box contained a small red velvet bag. I carefully removed it, peeked inside, and couldn't believe my eyes. The bag contained the initialled coin.

My grandfather's magic trick sparked a fascination with conjuring that has lasted throughout my life. In my teens I became one of the youngest members of a world-famous magic club, the Magic Circle. In my twenties, I worked as a professional magician, performing card tricks at some of

London's most fashionable West End restaurants. Once in a while, I even made an initialled coin disappear, and reappear in a little cloth bag sealed in a small tin box. Deceiving people on a twice-nightly basis sparked a strong sense of curiosity about why people are fooled. That interest acted as the catalyst for a psychology degree, and, some twenty years on, I still haven't shaken my fascination with the psychology of deception.

Over the years, I have unravelled the truth about deception, investigating the telltale signs that give away a liar, how fake smiles differ from genuine grins, and how people can be fooled into believing that they have experienced events that didn't actually happen.

We begin our journey into the shady world of skulduggery by examining an unusual body of work concerned with the evolutionary origins of deceit. It is a strange story involving a group of trunk-swinging elephants, talking apes, and children taking prohibited peeks at their favourite toys.

Jumbo deception, talking apes, and lying children

A few years ago, animal researcher Maxine Morris spotted some rather curious behaviour whilst observing a group of Asian elephants at Washington Park Zoo.[1]

At feeding time, each elephant was given a big bundle of hay. Morris noticed that a couple of the elephants tended to eat their own hay quickly, sidle up to their slower-eating companions, and then start swinging their trunks from side to side in a seemingly aimless way. To the uninformed, it appeared that these elephants were just passing the time of day. However, Morris's repeated observations suggested that this apparently innocent behaviour masked a duplicitous intent. Once the trunk-swinging elephants were sufficiently close to another elephant, they would suddenly grab some of the uneaten hay, and quickly gobble it up. Elephants are notoriously short-sighted, and so the slow-eating elephants were often completely unaware of the theft.

It is tempting to view these trunk-swinging/hay-stealing episodes as evidence of a carefully planned and executed deception. A kind of jumbo version of *Ocean's Eleven*. This may, however, be little more than wishful thinking. In the same way that we talk to our computers and cars as if they are people, so we have a tendency to humanize the behaviour of our four-legged friends. The seemingly deceitful elephants may simply have carried out the trunk-swinging/hay-stealing combination once by chance, rather liked the resulting excess of hay, and repeated the pattern without really thinking about it. The only way to know for certain would be to discover what was actually going on inside an elephant's head. The bad news is that elephants are in no position to describe their innermost thoughts and feelings. The good news is that some researchers believe that this has been achieved not with elephants, but with one of our closest evolutionary ancestors.

In the 1970s, talking gorillas were all the rage. As part of a large-scale research programme exploring interspecies communication, developmental psychologist Dr Francine Patterson from Stanford University attempted to teach two lowland gorillas called Michael and Koko a simplified version of American Sign Language.[2] According to Patterson, the great apes were capable of holding meaningful conversations, and could even reflect upon profound topics, such as love and death. Many aspects of the gorillas' inner lives appear remarkably similar to our own. Michael, for instance, liked watching the children's programme *Sesame Street*, whilst Koko preferred *Mister Roger's Neighborhood*. In 1998, Koko made a guest appearance on her favourite show, helping teach children that 'there is more to a person than what you see on the outside'. Michael likes painting, and has produced a large number of artworks, including self-portraits and several still- life representations. His work has proved remarkably popular with humans, and has been shown in various exhibitions. Koko is also no stranger to public attention. She has appeared in several films, and was the inspiration behind Amy, the talking ape, in Michael Crichton's bestseller, *Congo*. Koko also features in a promotional video on her website (using her communicative skills to appeal for donations) and, in 1998, took part in the first interspecies web chat. The opening lines of the conversation between the interviewer, Koko, and Dr Patterson illustrate some of the difficulties associated with trying to understand gorilla small talk:[3]

Interviewer: I'll start taking questions from the audience now. Our first question is: 'Koko, are you going to have a baby in the future?'

Koko: Pink.

Patterson: We've had earlier discussion about colours today.

Koko: Listen, Koko loves eat.

Interviewer: Me too!

Patterson: What about a baby? She's thinking . . .

Koko: Unattention.

Patterson: She covered her face with her hands . . . which
means it's not happening, basically, or it hasn't
happened yet.

Despite difficulties, Michael and Koko's trainers believe
that they have uncovered instances wherein their two hairy
colleagues were a tad economical with the truth.[4] In one
example, Koko broke a toy cat, and then signed to indicate
that the breakage had been caused by one of her trainers.
In another episode, Michael ripped a jacket belonging to
a trainer, and, when asked who was responsible for the
incident, signed 'Koko'. When the trainer expressed some
scepticism about his answer, Michael appeared to change
his mind, and indicated that Dr Patterson was actually to
blame. When the trainer pressed the issue again, Michael
finally looked sheepish (which isn't easy for a gorilla), and
then confessed all. Whereas the instances of alleged deception
among elephants were based purely upon observation, the
apes' apparent linguistic skills seem to provide much more
compelling evidence of intentional deceit.

The possibility of talking and lying apes has generated
fierce debate among researchers. Proponents claim that
Michael and Koko are clearly able to express their innermost
thoughts and emotions, and that the behaviour portrayed
during the 'Who ripped the jacket?', 'It was her' episodes are
clear evidence of deception. In response, critics argue that
the trainers are too eager to read meaning into the gorillas'
random actions, and that, when it comes to lying, the great
apes might simply be repeating behaviours that had got them
out of trouble before. As with the hay-stealing elephants, it is
almost impossible to know for certain.

Because of the difficulties of trying to decide if elephants and gorillas really are capable of lying, other researchers have explored the development of deception in the next best thing – children.

Some of the most interesting experiments examining children who cheat have involved asking youngsters not to take a peek at their favourite toys.[5] During these studies, a child is led into a laboratory and asked to face one of the walls. The experimenter then explains that he is going to set up an elaborate toy a few feet behind them. After setting up the toy, the experimenter explains that he has to leave the laboratory, and asks the child *not* to turn around and peek at the toy. The child is secretly filmed by hidden cameras for a few minutes, and then the experimenter returns and asks them whether they peeked. Almost all 3-year-olds do, and then half of them lie about it to the experimenter. By the time the children have reached the age of 5, all of them peek and all of them lie. The results provide compelling evidence that lying starts to emerge the moment we learn to speak. Perhaps surprisingly, when adults are shown films of their children denying that they peeked at the toy, they are unable to detect whether their darling offspring are lying or telling the truth.

Adhering to the old theatrical adage of never working with children or animals, my own research into lying has focused on adult deceit.

The language of lying

Lies have changed the course of world history. The lies told by Adolf Hitler to British prime minister Neville Chamberlain, when the two met just prior to the outbreak of war in September 1938, are famous. Hitler was secretly preparing to invade Czechoslovakia, and was therefore eager to prevent the Czechs from assembling a retaliatory force. The Führer assured Chamberlain that he had absolutely no intention of attacking Czechoslovakia, and the British leader believed him. A few days after their meeting, Chamberlain wrote to his sister, describing how he believed Hitler to be '. . . a man who could be relied upon when he had given his word'. Chamberlain was so convinced of Hitler's honesty that he urged the Czechs not to mobilize their troops, fearing that such a move might be viewed as an act of aggression by the Germans. The subsequent German attack quickly overwhelmed the ill-prepared Czechoslovakian forces, and led to the start of the Second World War. The world might now be a very different place had Chamberlain been able to detect Hitler's lies during their fateful meeting.[6]

World leaders are not the only people to lie, and be lied to. Deception affects each and every one of us. A few years ago I carried out a national survey into lying, in collaboration with the *Daily Telegraph*.[7] Only 8 per cent of respondents claimed never to have lied, and I suspect that most of these people couldn't bring themselves to tell the truth even in an anonymous survey. Other work has involved asking people to keep a detailed diary of every conversation that they have, and of all of the lies that they tell, over a two-week period. The results suggest that most people tell about two important lies each day, that a third of conversations involve some form of deception, that four in five lies remain undetected, that over 80 per cent of people have lied to secure a job (with most saying that they thought employers expected candidates to be dishonest about their background and experience), and that over 60 per cent of the population have cheated on their partners at least once.[8]

Are you a good liar? Most people think that they are, but, in reality, there are big differences in how well we can pull the wool over the eyes of others. There is, however, a very simple test that you can take to help determine your ability to lie. In fact, you have already taken it.[9]

At the start of the book I asked you to draw the letter 'Q' (for quirkology) on your forehead. If you didn't do it then, please do it now. Using the first finger of your dominant hand, simply draw a capital letter 'Q' on your forehead. Some people draw the letter 'Q' in such a way that they themselves can read it. That is, they place the tail of the Q on the right-hand side of their forehead. Other people draw the letter in a way that can be read by someone facing them, with the tail of the Q on the left side of their forehead. This quick test provides a rough measure of a concept known as 'self-monitoring'. High self-monitors tend to draw the letter Q in a way in which it could be seen by someone facing them. Low self-monitors tend to draw the letter Q in a way in which it could be read by themselves. What has all this to do with lying? High self-monitors tend to be concerned with how other people see them. They are happy being the centre of attention, can easily adapt their behaviour to suit the situation in which they find themselves, and are skilled at manipulating the way in which others see them. As a result, they tend to be good at lying. In contrast, low self-monitors come across as being the 'same person' in different situations. Their behaviour is guided more by their inner feelings and values, and they are less aware of their impact on those around them. They also tend to lie less in life, and so not be so skilled at deceit.

I have presented this fun test to groups of people for many years. Over time, I have noticed that there are a small number of people who, upon hearing what the test is all about, quickly convince themselves that they traced the letter Q in the opposite direction to the way they actually drew it. These people are able to ignore the evidence right in front of them, and instead twist the facts to fit the person they want to be. As a result, the test provides a rough indicator of how good you are at deceiving both yourself and others.

The vast majority of psychological work into deception has not focused on the types of people who are good and bad liars. Instead, it has concentrated on the art and science of lie detection. Can people detect deceit? What are the telltale signs that give away a lie? Is it possible to teach people to become better lie detectors?

Soon after I took up my position at the University of Hertfordshire in 1994, I received a rather curious email that had been sent out to academics across Britain. The email explained that, as part of a week-long national celebration of science, resources were available for a large-scale experiment in which members of the public could participate. The experiment would reach an audience of millions, because it would be conducted live on one of the BBC's flagship science programmes, *Tomorrow's World*. The email ended by asking academics to submit ideas for the experiment. I thought that it would be interesting to test the lie-detection skills of the entire country, and so suggested asking several politicians to lie or tell the truth on the programme, and have the public try to identify the lies. That way, I argued, it would be possible to determine scientifically which political party had the best liars. A few weeks later I was delighted to discover that my proposal had been chosen, and I started to fine-tune the study.

After a large number of telephone calls, the situation became clear – politicians were unwilling to participate, allegedly because they were terrible liars (none of us believed them). We looked for a prestigious alternative, and invited a legendary television political interviewer, Sir Robin Day, to be our guinea pig. Sir Robin was to the BBC what Walter Cronkite was to CBS. His penetrating and abrasive style of interviewing politicians had made him one of the most trusted figures on British television, and earned him the title of 'Grand Inquisitor'. We were delighted when Sir Robin accepted our challenge.

The design of the experiment was simple. I would interview Sir Robin twice, and in each interview ask him to describe his favourite film. In one interview he would say nothing but the truth, and in the other he would produce a complete pack of lies. We would then show both interviews on television, and see whether the public could detect which interview contained the lies.

The BBC assigned a talented young director named Simon Singh to the project. Simon went on to write several best-selling science books, including *Fermat's Last Theorem* and *The Code Book*. The two of us have worked on various projects together over the years, but first met to film Sir Robin's 'truth' and 'lying' interviews in the foyer of a large London hotel. Just after we had finished setting up the camera, the door swung open and in walked Sir Robin. His trademark thick-rimmed glasses and colourful bow tie made him instantly recognizable. As he sat down in front of the camera, he seemed slightly nervous that he was about to receive questions rather than ask them. I commenced the first of the two interviews, asking him to describe his favourite film. He explained that he had a great love for that Clark Gable classic, *Gone With The Wind*:

> **Richard Wiseman:** So, Sir Robin, what's your favourite film?
>
> **Sir Robin:** *Gone With The Wind*.
>
> **Richard Wiseman:** And why's that?
>
> **Sir Robin:** Oh, it's, it, it's a classic. Great characters; great film star – Clark Gable; a great actress – Vivien Leigh. Very moving.
>
> **Richard Wiseman:** And who's your favourite character in it?
>
> **Sir Robin:** Oh, Gable.
>
> **Richard Wiseman:** And how many times have you seen it?
>
> **Sir Robin:** Um . . . (*pause*) I think about half a dozen.
>
> **Richard Wiseman:** And when was the first time that you saw it?
>
> **Sir Robin:** When it first came out. I think that it was in 1939.

Once he had finished, I repeated the questions and he described being a big fan of the Marilyn Monroe movie, *Some Like It Hot*:

> **Richard Wiseman:** So, Sir Robin, what's your favourite film?
>
> **Sir Robin:** Ah . . . (*pause*) er, *Some Like It Hot*.
>
> **Richard Wiseman:** And why do you like that?

Sir Robin: Oh, because it gets funnier every time that I see it. There are all sorts of bits in it which I love. And I like them more each time that I see it.

Richard Wiseman: Who's your favourite character in it?

Sir Robin: Oh, Tony Curtis, I think. He's so pretty . . . (*short pause*) and he's so witty, and he mimics Cary Grant so well and he's very funny the way he tries to resist being seduced by Marilyn Monroe.

Richard Wiseman: And when was the first time that you saw it?

Sir Robin: I think when it came out, and I forget when that was.

Which do you think is the lie?

The full experiment took place a few weeks later on a live edition of *Tomorrow's World*. At the start of the programme, we played the two interviews and asked viewers to decide which they thought was the lie, and register their vote by telephoning one of two numbers. It was the first time that anything like this had been attempted, and Simon and I had absolutely no idea whether people would make a telephone call in the name of science. We needn't have worried. Within minutes we had received over 30,000 calls.

When the lines closed, we quickly set about analysing our results. Fifty-two per cent of viewers thought that Sir Robin had been lying about *Gone With The Wind*, and 48 per cent had voted for *Some Like It Hot*. We then showed viewers a short film clip of me asking Sir Robin whether he really liked *Gone With The Wind*. His reply was short and to the point: 'Good heavens no! It's the most crashing bore. I fall asleep every time I see it.' At the end of the programme we announced the findings, and explained that when it came to detecting deceit, the public's skills were little better than chance.[10]

It could be argued that Sir Robin was an extremely skilled liar, and that in everyday life people are much better at detecting deception. To investigate this, one would have to carry out lots of experiments using many different types of people, who lie and tell the truth about a huge range of issues. It would be a mammoth task, but for the last thirty years or so, this is exactly what a small band of highly dedicated psychologists have been doing.[11] They have had people visit art exhibitions and lie about their favourite paintings, steal money from a wallet and then deny the theft, endorse products that they dislike intensely, and watch films depicting amputations whilst trying to convince others that they are looking at a relaxing beach scene. The research has studied the lying behaviour of salespeople, shoppers, students, drug addicts, and criminals. Some of my work in this area has involved showing people videotapes of instances in which people have made high-profile public appeals for information about a murder, only later to confess and be convicted of the crime themselves.

The results have been remarkably consistent – when it comes to lie detection, the public might as well simply toss a coin. It doesn't matter if you are male or female, young or old; very few people are able to reliably detect deception. The results suggest that we can't even tell when our partners are being economical with the truth.[12] In a series of experiments exploring romantic deception, one member of a long-term couple was presented with a series of slides containing images of a highly attractive person of the opposite sex, and asked to try to convince their partner that they found the attractive person unattractive. The findings suggest that most people in long-term relationships are dreadful at telling when their partner is lying. Some researchers believe that many long-term couples have remained together precisely because they cannot spot one another's lies.

Perhaps the public shouldn't worry too much about their inability to detect lies. They are, after all, in good company. Psychologist Paul Ekman from the University of California, San Francisco, showed videotapes of liars and truth-tellers to various groups of experts, including polygraph operators, robbery investigators, judges, and psychiatrists, and asked them to try to identify the lies.[13] All tried their best. None of the groups performed better than chance.

So why are people so bad at detecting deceit? The answer lies in the work of psychologists like Professor Charles Bond from the Texas Christian University.[14] Bond has conducted surveys into the sorts of behaviours that people associate with lying. Unlike some psychological research, his work doesn't involve asking a few hundred American university students to tick boxes on a form. Instead, he has surveyed thousands of people from over sixty countries, asking them to describe how they go about telling whether someone is lying. People's answers are remarkably consistent. From Algeria to Argentina, Germany to Ghana, Pakistan to Paraguay, almost everyone thinks that liars tend to avert their gaze, nervously wave their hands around, and shift about in their seats.

There is, however, one small problem. Researchers have spent hours upon hours carefully comparing films of liars and truth-tellers. The work involves trained observers sitting in front of a computer watching digitized videos again and again. On each showing, the observers look out for a particular behaviour, such as a smile, blink, or hand movement. Each time they see what they are looking for, they press a button and the computer records their response. Each minute of footage takes about an hour to analyse, but the resulting data allows researchers to compare the behaviour associated with a lie and truth, and thus uncover even the subtlest of differences. The results are clear. Liars are just as likely to look you in the eye as truth-tellers, they don't move their hands around nervously, and they don't shift about in their seats (if anything, they are a little more static than truth-tellers). People fail to detect lies because they are basing their opinions on behaviours that are not actually associated with deception.

So, what are the signals that really give away a liar? To answer the question, researchers have searched for reliable differences between the behaviour of liars and truth-tellers. The answer, it seems, lies in the words we use and the way in which we say them.[15] When it comes to lying, the more information you give away, the greater the chances of some of it coming back to haunt you. As a result, liars tend to say less, and provide fewer details, than truth-tellers. Look back at the transcripts of the interviews with Sir Robin. His lie about *Gone With The Wind* contains about forty words, whereas the truth about *Some Like It Hot* is nearly twice as long. Now take a look at the level of detail in the two interviews. In the first interview, he presents a very general description of the film, merely stating that it is a classic with great characters. When he tells us the truth, however, he provides far more detail, describing a scene in which actor Tony Curtis tries to resist being seduced by Marilyn Monroe.

When it comes to the language of lying, this is just the tip of the iceberg. Liars often try to psychologically distance themselves from their falsehoods, and so tend to include fewer references to themselves, and their feelings, in their stories. Again, Sir Robin's testimony provides a striking illustration of the effect. When he lies, Sir Robin mentions the word 'I' just twice, whereas when he tells the truth his account contains seven 'I's. In his entire interview about *Gone With The Wind*, Sir Robin only once mentions how the film makes him feel ('Very *moving*'), compared to the several references to his feelings when he talks about *Some Like It Hot* ('it gets *funnier* every time I see it', 'all sorts of bits I *love*', '[Curtis is] so *pretty* . . . so *witty*').

Then there is the issue of forgetting. Imagine someone asking you a series of questions about what you did last week. It is quite probable that you wouldn't be able to remember many of the trivial details and, being the honest person you are, would admit to your memory lapse. Liars tend not to do this. When it comes to relatively unimportant information, they seem to develop superpowered memories and often recall the smallest of details. In contrast, truth-tellers know that they have forgotten certain details, and are happy to admit it. Sir Robin's interviews illustrate the point. There is only one instance in the interviews when he admits that he cannot remember a detail, and it is when he tells us the truth about not being able to recall the first time he saw his favourite film, *Some Like It Hot*.

Research has yet to confirm exactly why body language is often misleading, whilst the language of lying is so revealing. One theory is that eye contact, and hand movements, are easy to control and so liars can use these signals to convey whatever impression they want. In contrast, trying to control the words we use and the way we say them is much harder, and so a person's use of language becomes a far more reliable guide to the truth.

Whatever the theory, the simple fact is that the real clues to deceit are in the words that people use. So do people become much better lie detectors when they listen to a liar, or even just read a transcript of their comments? I have to own up to a little falsehood of my own. I didn't tell you the whole truth about the experiment with Sir Robin. Like all good deceivers, I didn't actually lie to you, I just left out some important information.

Leslie Nielsen, ketchup, and sour cream

The television experiment was just one small part of a much bigger study. On the same day as the BBC programme was aired, we also played just the soundtrack of the two interviews on a national radio station, and science editor Roger Highfield arranged for the transcripts to be printed in the *Daily Telegraph*. Each time, listeners and readers were asked to guess which interview they thought contained the lies, and register their opinion by telephoning one of two numbers. Thousands of people were kind enough to participate. Although the lie-detecting abilities of the television viewers were no better than chance, the newspaper readers were correct 64 per cent of the time, and the radio listeners scored an impressive 73 per cent accuracy rate. When it comes to detecting lies, people are better off listening rather than looking.

The experiment with Sir Robin is far from being the only study to illustrate that people's lie-detecting abilities are increased by encouraging them to listen, rather than look. One of the more unusual pieces of research in the area was carried out by Glenn Littlepage and Tony Pineault from Middle Tennessee State University.[16] These researchers carried out their study using one of the best-known, and longest-running, game shows on American television. *To Tell the Truth* involved three contestants each claiming to be the same person. This trio was interrogated by four celebrity panellists who tried their best to uncover who was genuine and who was bluffing. After they had made their decisions, the host asked the truth-teller to stand up and reveal all. This show has become a part of American popular culture, and formed the basis for the opening sequence of the film *Catch Me If You Can*. Littlepage and Pineault taped various editions of the show. In one of their episodes three women claimed to be an expert on the Middle Ages, and in another three men each said that they had been asked by the People's Republic of China to discover the remains of a prehistoric 'Peking Man'. The researchers then showed the clips to various groups of people. One of the groups watched the show as normal, with both the sound and images. Another group heard only the soundtrack of the shows, whilst a third group saw only the images. The results demonstrated the importance of the language of lying. Those who saw just the images were terrible at spotting the bluffing contestant, whereas those who heard just the sound-track were surprisingly skilled at working out who was about to stand up and be revealed as the truth-teller.

It is time to test your new lie-detection skills. A few years ago, a science show called *The Daily Planet* on the Canadian

Discovery Channel asked me to help carry out another national lie detection experiment. They persuaded one of my childhood heroes, Hollywood actor and comedian Leslie Nielsen (star of *Airplane*, *Naked Gun*, and *Police Squad!*), to be our guinea pig. Nielsen was interviewed twice by the show's host, Jay Ingram. In each interview Nielsen was asked about his favourite food. As with the experiment with Sir Robin Day, one of the answers was a complete pack of lies and the other was the honest truth. Can you spot the lie this time?

Interview 1

Jay Ingram: What is your favourite food?

Leslie Nielsen: What is my favourite food? What is my favourite food? And I can take my pick out of absolutely anything? Humm . . . boy, that's a toughie, I tell you. It really depends. I guess . . . my favourite food is ketchup.

Jay Ingram: Ketchup! Why do you like ketchup so much?

Leslie Nielsen: I don't know. I think I am one of those people who is capable of putting ketchup on absolutely anything, or everything, whichever way you want to look at it. Yes, ketchup.

    I am thinking really mainly about something that is a holdover from the time when I was a little boy. You know, how you go looking for it – you say, 'Hey, Mommy, give me a piece of bread and jam.' And I remember the time when my mother, she said, 'We don't have any jam, Leslie, we don't have any jam.' I said, 'But, but, but.' And she said, 'I'll give you something.' And she had a piece of bread and butter, and she put ketchup on it. And smoothed it over and so on. I'm addicted to it, and, I know I would catch myself, when I was . . . if I was feeling good around the house, no matter what it is, and I got hungry, I would head for the refrigerator and get out a piece of bread and butter and put ketchup on it. It made me feel even better.

Interview 2

Jay Ingram: So, Leslie Nielsen, what is your favourite food?
Leslie Nielsen: It's becoming a favourite for me . . . it's at the head of my list . . . I'm really only going by what comes into my mind first. And . . . you know sour cream. You take a dollop of sour cream and you put it on guacamole, for example, or . . . I think it is because I have got into a Mexican tinge here, and I can remember my mother, for example, when I was a kid, she would eat a tomato sandwich with mayonnaise on it. Well mayonnaise, later on, it looked like sour cream, it would be the last thing in the world that I would want to touch.

And . . . errr . . . so I really stayed away from it, but today . . . it's a very unusual flavour, and you can get it more or less low fat which I am very careful and cautious about, and it is a new taste for me, but is something that I am growing very rapidly to like very much – sour cream.

As you might already have guessed, Leslie loves ketchup and hates sour cream. The transcripts contain the linguistic patterns that are typical of lying and truth-telling. First, the lie is far shorter than the truth – Leslie used about 220 words when he spoke about ketchup and roughly 150 when he described his 'love' of sour cream. The transcripts also contain evidence of the 'psychological distancing' associated with lying. When Leslie tells the truth, he uses the word 'I' seventeen times, compared to just nine times when he lies. Also, the truth contains a fairly detailed description of the type of childhood experience that he associates with ketchup, with several descriptions of his emotions ('I'm *addicted* to it', 'if I was *feeling good*', and 'it made me *feel even better*'). In contrast, Leslie's lie is far more factual (about how you can use it, that it has an unusual flavour, and is low fat), and only contains a single and slightly strange reference, right at the very end of the interview, to how it makes him feel ('it . . . is something that I am growing very rapidly to like very much').

Once you know the telltale signs associated with the language of a lie, detecting deceit becomes much easier. It has little to do with whether liars look you in the eye, move their hands around, or shift about in their seat. The most reliable signs of lying are in a person's voice, and their unconscious choice of language. The lack of key details in their descriptions; the increase in pauses and hesitations; the way liars distance themselves from their deceit by avoid-ing self-references such as 'I' and failing to describe their feelings. The way they seem to be able to remember minute information that truth-tellers forget. Learn to listen for the secret signals and the thin veil of deception is lifted. Suddenly you see what people really think and feel, and the world becomes a very different place. Honestly. Trust me on this one.

The *Mona Lisa*, freshly guillotined heads, and the School Sisters of Notre Dame

So do the results of the studies conducted with Sir Robin Day and Leslie Nielsen mean there are no telltale signs of deception that can be detected in people's body language and facial expressions? Not exactly. In fact, there are ways of spotting deception using your eyes rather than your ears, it's just that you have to know exactly what you are looking for. Let's consider one of the most common, and frequently faked, forms of non-verbal behaviour – the human smile.

We all smile, but few of us have any insight into the complex psychology underlying this seemingly simple behaviour. Do you smile because *you* are happy, or to let *other people* know that you are happy? This seemingly simple question has generated fierce debate among researchers. Some have argued that the smile is driven almost entirely by an inner feeling of happiness, whilst others have claimed it is a social signal designed to let those around you know how you feel. To help settle the issue, Professors Robert Kraut and Robert Johnston from Cornell University decided to compare the amount people smiled when they were happy but alone, with when they were equally happy but with others.[17] After much deliberation, they hit upon the perfect place to conduct their study – a bowling alley. They realized that when bowlers rolled their ball down the lane and obtained a good score, they tended to be alone and happy. When they turned and faced their fellow bowlers, however, they would be just as happy but now interacting with others.

Over the course of several studies, Kraut and his colleagues secretly observed over 2000 bowls. Each time, the researchers carefully documented the course of events, including the bowler's facial expressions, their score, and whether they were facing the bowling lane or their friends. In one part of the study, the team quietly muttered the information into a dictaphone (avoiding possible suspicion by using code words for each factor) to ensure that their measurements were accurately recorded. Their findings revealed that only 4 per cent of bowlers smiled when they obtained a good score, but were facing away from their colleagues. However, when the bowlers turned around and looked at their friends, 42 per cent of them had a huge smile pasted across their faces. Strong evidence that we don't smile simply because we are happy, but rather to let others know that we are happy.

Like all social signals, smiling is open to fakery. People often smile to give the impression that they are happy when, deep down, they feel less than joyous. But is a fake smile identical to a genuine smile, or are there telltale facial signals that separate the two expressions? It is an issue that has taxed researchers for over a hundred years, and one that lay at the heart of an unusual experiment that I recently conducted in an art gallery.

In the previous chapter I described how the New Zealand Science Festival were kind enough to let me undertake the second part of the Born Lucky experiment. Before the trip, I suggested to the festival organizers that we stage a second study that would help unmask the secrets of the fake smile. The idea was simple. I wanted people to see several pairs of photographs.[18] Each pair would contain two images of the same person smiling. One of the smiles would be genuine and the other fake, and the public would be asked to spot the genuine grins. A careful comparison of the photographs would reveal whether there were any telltale signs that give away a fake smile, and an analysis of results would determine whether people were able to utilize these signals. After some discussion, we hit upon the idea of staging the study as an exhibition in an art gallery. The Dunedin Public Art Gallery kindly agreed to host the event, ensuring that our unusual art–science exhibition would brush shoulders with works by Turner, Gainsborough, and Monet.

For my smiling experiment I needed to create a way of having people produce both a genuine, and a fake, smile. Researchers have used a diverse range of techniques to provoke such facial expressions in the laboratory. In the 1930s, psychologist Carney Landis wanted to photograph people's faces as they experienced a range of emotions, and so asked volunteers to listen to a jazz record, read the Bible, and flick through some pornographic images (Landis reports that 'the experimenter was careful not to laugh or appear self-conscious during this latter situation').[19] Two other situations were designed to provoke more extreme reactions. In one, volunteers were asked to place their hand in a pail of water containing three live frogs. After the volunteers had reacted, they were urged to continue feeling around in the water. The experimenter then ran a high voltage through the water, giving them a considerable electric shock. However, Landis's pièce de résistance involved his final, and ethically most questionable, task. Here participants were handed a live white rat and a butcher's knife, and asked to behead the rat. Around 70 per cent of the volunteers carried out the task after 'more or less urging', and in the remaining cases the experimenter beheaded the rat himself. Landis reports that 52 per cent of people smiled during the beheading, compared to 74 per cent receiving the electric shock. Most of the volunteers were adults, but the group did include a 13-year-old boy who was a patient at the University Hospital because he was emotionally unstable and had high blood pressure ('So, son, what happened at the hospital today?').

In my smiling study, I asked each of our volunteers to bring along a poodle, and a large knife. Just kidding. Actually, we opted for two tasks that were a little less controversial. Each person was asked to come along with a friend. Whenever their friend made them laugh, we took a photograph of their genuine smile. We then asked them to imagine that they had just met someone they disliked, and had to fake a polite smile. Two of the resulting images are shown below. This pair of photographs, and nine other pairs, formed the basis for the exhibition.

I am not the first academic to examine the science of smiling in an art gallery. In 2003, Harvard neuroscientist Professor Margaret Livingstone attempted to solve scientifically the mystery of perhaps the most famous smile in art.[20] The *Mona Lisa* was created by Leonardo da Vinci in the 1500s, and has perplexed art historians for hundreds of years. Much of the debate has revolved around the nature of her enigmatic facial expression, with some scholars arguing that the painting clearly shows a smiling face, whilst others claim that the expression is one of great sadness. In 1852, a young French artist threw himself from the fourth-floor window of his Paris hotel after writing, 'For years I have grappled desperately with her smile. Now I prefer to die.' Professor Livingstone took a somewhat more constructive approach to the issue.

For years, people had noticed that the *Mona Lisa*'s smile was far more apparent to people when they looked at her eyes, and appeared to vanish when they looked directly at her mouth. This was clearly a key part of the enigmatic nature of the painting, but people couldn't work out how Leonardo had created the strange effect. Professor Livingstone discovered that the illusion was due to the fact that the human eye sees the world in two very different ways. When people look at something directly, the light falls on a central part of their retina called the fovea. This part of the eye is excellent at seeing relatively bright objects, such as those in direct sunlight. In contrast, when people see something out of the corner of their eye, the light falls on the peripheral part of the retina, which is much better at seeing in semi-darkness. Livingstone found that Leonardo's picture is using the two parts of the retina to fool people's eyes. Analyses showed that the great artist had cleverly used the shadows from the *Mona Lisa*'s cheekbones to make her mouth much darker than the rest of her face. As a result, the *Mona Lisa*'s smile appears more obvious when people look at her eyes because they are seeing it in their peripheral vision. When people look directly at her mouth, they are seeing the dark area of the painting more clearly with their fovea, and so the smile looks far less pronounced.

Livingstone was not the first scientist to be intrigued with the mystery of the human smile. Two hundred years before, a small band of European scientists conducted a series of strange studies into exactly the same topic.

In the early nineteenth century, researchers were fascinated by how electricity might be used to gain important anatomical and physiological insights. Some of this work involved the rather gruesome, and often public, stimulation of fresh corpses. Perhaps the most famous exponent of this technique was the Italian scientist Giovanni Aldini.[21] Aldini specialized in bringing murderers back to life. In one well-publicized example of his work, Aldini travelled to London to reanimate a murderer named George Foster. Foster had been found guilty of killing his wife and child by drowning them in a canal, and had been sentenced to hang on 18 January 1803. Shortly after his death, Foster's body was taken to a nearby house, where Aldini applied various voltages to his body under the watchful eye of eminent British scientists. The court calendar (appearing, rather appropriately, in an edited collection of papers by a Mr G. T. Crook) reported the results:[22]

On the first application of the process to the face, the jaws of the deceased criminal began to quiver, and the adjoining muscles were horribly contorted, and one eye was actually opened. In the subsequent part of the process the right hand was raised and clenched, and the legs and thighs were set in motion. It appeared to the uninformed part of the by-standers as if the wretched man was on the eve of being restored to life.

The description then went on to rule out the possibility of resurrection, on the grounds that during Foster's hanging, 'several of his friends who were under the scaffold had violently pulled his legs, in order to put a more speedy termination to his suffering', and noted that even if Aldini's demonstration had brought Foster back to life, he would have been hanged a second time, as the law demanded that such criminals be 'hanged until they were dead'. The calendar also describes how one of the onlookers, a Mr Pass, the beadle of the Surgeons' Company, was so shocked by the demonstration that he died of fright soon after his return home, thus making Foster one of the very few murderers to kill another victim *after* his own death.

Aldini was not the only scientist to experiment with the effects of electrical stimulation upon the muscles of the body. A few years later, a group of Scottish scientists carried out a similar demonstration with another murderer, contorting the man's face into 'fearful action', and setting his fingers in motion such that 'the corpse seemed to point to different spectators'. Again, the resulting scene proved too much for many onlookers, with one man fainting and several leaving the room in disgust.[23]

In addition to laying the foundations for much modern-day work into the effects of medical electrical stimulation, this research resulted in two significant contributions to popular culture. The use of electricity apparently to resurrect the dead helped inspire Mary Shelley to write *Frankenstein*. Also, the word 'corpsing' – a term used by actors when they suddenly laugh while trying to be serious – originates in the inappropriate grins exhibited on the lifeless heads.

Aldini's work also inspired a French scientist named Guillaume Duchenne de Boulogne to develop a more sophisticated system for investigating which muscles are involved in different facial expressions. Rather than work with recently executed murderers, Duchenne decided to take the somewhat more civilized approach of photographing living subjects as electricity was applied directly to their faces. After much searching, he found a participant who was willing to put up with the constant, and rather painful, stimulation of his face. In his 1862 book, *The Mechanism of Human Facial Expression*,[24] Duchenne presents a less than flattering description of his guinea pig:

> The individual I chose as my principal subject for the experiments . . . was an old toothless man, with a thin face, whose features, without being absolutely ugly, approached ordinary triviality, and whose facial expression was in perfect agreement with his inoffensive character and his restricted intelligence.

In addition, the man had another desirable attribute – almost total facial anaesthesia. This meant that Duchenne could '. . . stimulate his individual muscles with as much precision and accuracy as if I were working with a still irritable cadaver' (see opposite).

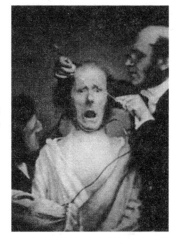

Duchenne stimulates the face
of his volunteer

After snapping hundreds of photographs, Duchenne uncovered the secret of the fake smile. When electricity was applied to the cheeks of the face, the large muscles on either side of the mouth – known as the zygomatic major – pulled the corners of the lip upwards to create a grin. Duchenne then compared this smile with the one produced when he told his thin-faced participant a joke. The genuine smiles included not only the zygomatic major, but also involved the orbicularis oculi muscles that run right around each eye. In a genuine smile these muscles tighten, pulling the eyebrows down and the cheeks up, producing tiny crinkles around the corners of the eyes. Duchenne discovered that the tightening of the eye muscles lay outside of voluntary control, and was 'only put into play by the sweet emotions of the soul'.

Duchenne's work has been confirmed by more recent research,[25] and our twenty-first-century images of genuine and false smiles showed exactly the same effect. Take another look at the two photographs on page 66. The image on the right shows a fake smile, in which the zygomatic major muscles are pulling up the corners of the mouth. Researchers have recently christened this the 'Pan American' smile, after the fake grin often produced by flight attendants working for the now-defunct airline. The left-hand picture shows a genuine smile, involving both the zygomatic major muscles, and the orbicularis oculi muscles around the eyes. Here, the upward movement of the cheek has produced more striking contour lines at the sides of the nose, and around the bottom and sides of the eyes. Also, the eyebrows, and the skin below the eyebrows, have moved down towards the eyes, making them slightly narrower and creating a 'bagging' directly above the eyes. These subtle changes are easier to see in the enlargements of the photographs shown below (the genuine smile is shown in the top image, and the fake in the bottom).

Hundreds of people visited the Dunedin Art Gallery during the festival and were kind enough to take part in the experiment. Participants were given a questionnaire, asked to look at each pair of photographs, and indicate which of them they believed showed the genuine smile. The results revealed that many people couldn't tell the fake from the genuine smiles, and even those that thought they were especially sensitive to the emotions of others scored little better than chance. Had they known exactly what to look for, however, the telltale clue was right in front of their noses (and just to the sides of the noses of the people in the photographs).

The people participating in the study were not especially skilled at spotting a genuine smile. The ability to differentiate between the two, however, using the type of system developed by Duchenne, has provided psychologists with a unique insight into the relationship between emotion and everyday life, to the point that researchers have recently started to look at the way small and seemingly irrelevant aspects of a person's behaviour in early life may provide a useful insight into their long-term success and happiness.

The idea is beautifully illustrated by a study involving 200 nuns carried out by psychologist Deborah Danner at the University of Kentucky.[26] Prior to joining an American nunnery known as the School Sisters of Notre Dame, each nun is required to write an autobiographical account of her life. In the early 1990s, Danner analysed 180 autobiographies written by nuns who had joined the order in the mid 1970s, counting the frequency with which words describing positive emotions, such as 'joy', 'love', and 'content', appeared. Remarkably, nuns who described experiencing a large number of positive emotions lived as much as ten years longer than the others.

Similar work suggests that the presence of the Duchenne smile in early adulthood provides a deep insight into people's lives. In the late 1950s, about 150 senior students at Mills College (a private women's college in Oakland, California) agreed to allow scientists to conduct a long-term study of their lives. During the course of the following fifty years, these women have provided researchers with ongoing reports about their health, marriage, family life, careers and happiness. A few years ago, Dacher Keltner and LeeAnne Harker from the University of California at Berkeley looked at the photographs of the women that had been taken for the college yearbook when they were in their early twenties.[27] Nearly all of the women were smiling. However, when the researchers carefully examined the images, they noticed that about half of the photographs showed a Pan Am smile and half a genuine Duchenne smile. They then went back to the information that had been provided by the women throughout their lives, and discovered something remarkable. Compared to the women with the Pan Am smiles, those displaying the Duchenne smiles were significantly more likely to be married, stay married, be happier, and be healthier throughout their lives.

Lifelong success and happiness can be predicted by the simple crinkling around the sides of the eyes that first caught Duchenne's attention over a century ago. Interestingly, Duchenne realized the importance of his discovery long before his fellow scientists. Summing up his feelings about his research at the end of his career, Duchenne noted:

You cannot exaggerate the significance of the fake smile.

The expression which can be both a simple smile of politeness, or act as a cover to treason. The smile that plays upon just the lips when our soul is sad.

'Never mind, son, we'll ride it down together'

When it comes to everyday deception, the language of lying and fake smiles is just the tip of the iceberg.

In the mid 1970s, psychologists started to take a serious look at the malleability of memory. In a classic series of experiments conducted by psychologist Elizabeth Loftus and her colleagues, participants were shown slides depicting a car accident.[28] Everyone saw a red Datsun driving along a road, turning at a junction, and then hitting a pedestrian. After seeing the slides, participants were fed misleading information in a very sneaky way. In reality, the slide of the road junction had contained a *stop* sign. However, the experimenters wanted to subtly deceive participants by suggesting that they had seen a different sign, and so asked them to name the colour of the car that drove through the *give way* sign. Later on, the participants were shown a slide of the junction showing either a *stop* or *give way* sign, and asked to say which they had seen before. The majority were sure they had originally seen a *give way* sign at the junction. The study led to a whole raft of similar experiments, resulting in people being persuaded to recall hammers as screwdrivers, *Vogue* magazine as *Mademoiselle* magazine, a clean-shaven man as having a moustache, and Mickey Mouse as Minnie Mouse.

Subsequent research revealed that the same idea can also

be used to deceive people into remembering events that haven't actually happened. One recent study, conducted by Kimberley Wade from the Victoria University of Wellington and her colleagues, demonstrated the power of the effect.[29] Wade asked twenty people to persuade a family member to participate in an experiment that was allegedly concerned with why people reminisce about childhood events. The experimenters also asked their recruiters to supply them secretly with a photograph of this person as a young child. The experimenters then used this image to create a fake photograph depicting a childhood trip in a hot-air balloon. One of the original photographs, and the resulting manipulated image, is shown below. Finally, the experimenters asked their recruiters to supply three other photographs showing the person taking part in various genuine childhood events, such as a birthday party, seaside visit, or day trip to the zoo.

The participants were interviewed three times over the course of two weeks. During each interview, they were shown the three genuine photographs and the fake one, and encouraged to describe as much about each experience as possible.

Genuine and doctored photographs used in Kimberley Wade's false memory experiments

In the first interview almost everyone was able to remember details of the genuine events, but about a third also said that they remembered the non-existent balloon trip, which some actually described in considerable detail. The experimenters asked all of the participants to go away and think more about the experiences. By the third and final interview, half of the participants remembered the fictitious balloon trip, and many described the event in some detail. One participant who, at an initial interview, correctly stated that they had never been in a hot-air balloon, ended up producing the following account of the non-existent event:

> I'm pretty certain it occurred when I was in form one at the local school there . . . basically for $10 or something you could go up in a hot-air balloon and go up about 20 odd meters . . . it would have been a Saturday and . . . I'm pretty certain that mum is down on the ground taking a photo.

Wade's work is just one of a long line of experiments show-ing that people can be manipulated into recalling events that simply didn't happen. In another set of studies, a group of participants were persuaded to provide a detailed account of how, as a child, they visited Disneyland and met someone dressed in a Bugs Bunny costume (Bugs is not a Disney char-acter, and so would not have been in Disneyland).[30] Then there was the time when experimenters interviewed the parents of potential participants, asking them whether their offspring had ever become lost in a shopping mall as a child.[31] After carefully selecting a group of people who had not expe-rienced this event, the experimenters managed to persuade a large percentage of them to provide a detailed account of this traumatic, but non-existent, experience. Related work has convinced people that they once experienced an overnight hospitalization for a high fever with a possible ear infection, accidentally spilled a punch bowl over the parents of the bride at a wedding reception, had to evacuate a grocery store because the sprinkler system was set off, and caused a car to roll into another vehicle by releasing its brake.[32] The work shows that our memories are far more malleable than we would like to believe. Once an authority figure suggests that we have experienced an event, most of us find it difficult to deny, and start to fill in the gaps from our imagination. After a while, it becomes almost impossible to separate fact from fiction, and we start to believe the lie. The effect is so powerful that sometimes it doesn't even require the voice of authority to fool us. Sometimes we are perfectly capable of fooling ourselves.

In December 1983, American President Ronald Reagan addressed the Congressional Medal of Honor Society. He decided to relate an alleged real-life story that he had told many times before.

Reagan described how, during the Second World War, a B-17 bomber was limping its way across the English Channel, badly damaged by anti-aircraft fire. The gun turret that hung beneath the plane had been hit, injuring the gunner inside, and jamming the door of the turret shut. The plane started to lose altitude and the commander ordered his men to bail out. The gunner was trapped in the turret, and knew he was going to go down with the plane. The last man to jump out of the plane later described how he saw his commander sitting next to the turret, telling the terrified gunner, 'Never mind, son, we'll ride it down together.'

Reagan explained how this remarkable act of courage had resulted in the commander being posthumously awarded the Congressional Medal of Honor, and ended this part of his emotional speech by noting how America had been right to award its highest honor '. . . to a man who would sacrifice his life simply to bring comfort to a boy who had to die'. It is a wonderful story, and suffers from only one small problem. It never actually happened. After checking the citations of all 434 Congressional Medals of Honor awarded during the Second World War, journalists could find no mention of the episode or anything like it. Eventually, a member of the public pointed out that the story was a near-perfect fit to events portrayed in the popular wartime film, *A Wing and a Prayer*. In the climactic scene of the film, a radio operator informs the pilot that the plane is badly damaged, and that he is injured and cannot move. The pilot replies, 'I haven't got the altitude, Mike. We'll take this ride together.'

The deceptive effects of suggestion do not stop at world leaders remembering fiction as fact. Exactly the same techniques are frequently used by professional deceivers to persuade people that they have experienced the impossible.

Remembering the impossible

Magicians are honest deceivers. Unlike most liars, they are completely open about the fact that they are going to cheat. Despite this, they still have to convince an audience that objects can disappear into thin air, that women can be sawn in half, and that the future can be predicted with uncanny accuracy.

For over a hundred years, a handful of psychologists have investigated the secret psychology used by magicians to fool their audiences. In the 1890s, American psychologist Joseph Jastrow teamed up with two world-famous illusionists to discover whether the hand really is quicker than the eye. Jastrow is one of my academic heroes. He was an amazing character who conducted many unusual investigations, including one of the first experiments into subliminal perception, analysing the dreams of blind people, and figuring out the psychology behind the Ouija board. Unfortunately, Jastrow also suffered from depression, with one Chicago newspaper reporting the onset of his illness with the headline, 'Famous mind doctor loses his own'.

To investigate the psychology of magic, Jastrow collaborated with two illusionists, named Alexander Herrmann and Harry Kellar.[33] Herrmann and Kellar were two of the most famous magicians of their day, and were locked in a constant battle of professional rivalry throughout most of their professional lives. If one made a donkey disappear, the other would make an elephant vanish. If one made a woman levitate above the stage, the other would have their assistant float a few feet higher. If one plucked a fan of cards from thin air, the other would perform the same feat blindfolded. Jastrow invited these two great performers to his University of Wisconsin laboratory, and had them participate in a range of tests measuring their reaction time, speed of movement, and accuracy of finger motion. Jastrow's results revealed little out of the ordinary, with the performers' data roughly matching those of a control group of non-magicians collected a few years before.

But Jastrow scientifically demonstrated what most magicians already knew. Magic has little to do with fast movements. Instead, conjurors use a range of psychological weapons to fool their audiences. Suggestion plays a key role in the process. In the same way that people can be made to believe that they once went on a non-existent trip in a hot air balloon, or became lost in a shopping mall, so magicians have to be able to manipulate people's perception of a performance.

The concept can be illustrated with a simple laboratory-based experiment that I recently conducted into mind over matter.[34] I showed a group of my students a videotape in which a magician apparently used the power of his mind (actually sleight of hand) to bend a metal key. He then placed the key on the table, stood back, and said, 'Look, it's amazing, the key is still bending.' Afterwards, all the students were interviewed about what they had seen. Over half were convinced that they had seen the key continuing to bend as it lay on the table, and had no idea how the magician could have achieved such an impressive trick: a dramatic illustration of how an expert deceiver can draw upon years of experience to deliver a sentence with such confidence that people believe they see the impossible happening before their very eyes.

Psychology in the seance room

Perhaps my most memorable set of studies examined the role of suggestion in the seance room.[35] Much of this work was carried out with a friend of mine, Andy Nyman. Andy is a skilled actor and magician, and helps create material for the highly successful British television illusionist, Derren Brown.

I first met Andy many years ago at a conference on magic, and we discovered that we were both interested in the techniques used by fraudulent mediums in the nineteenth century to fake ghostly phenomena in the seance room. We were curious to discover whether the hundred-year-old techniques would still fool a modern-day audience, and so decided to stage a series of unusual experiments.

The plan was simple. We would invite groups of people to attend a theatrical reconstruction of a Victorian seance, and use various techniques, including suggestion, to fake spirit activity. We would then ask them to tell us what they had experienced so that we could assess whether they had been fooled by our attempts at deception.

First we needed a spooky-looking venue. We came across the House of Detention – a dark, dank, disused, underground Victorian prison in the heart of London. It was perfect. The owners kindly allowed us to hire this uninviting venue for a week, and we staged two fake shows per evening, with twenty-five people attending each seance.

When people arrived they were asked to complete a short form asking them whether they believed in the existence of genuine paranormal phenomena. I then led the group through the maze of underground prison corridors, briefly relating the history of the Victorian seance. Eventually they were taken along a narrow ventilation shaft into a large room at the heart of the prison. Here Andy introduced himself to the group, and explained that he would be playing the part of the medium for the evening. Lit only by candlelight, he asked everyone to join him around a large table in the centre of the room.

For the next twenty minutes, Andy told the group a fictitious ghost story concerning the murder of a non-existent Victorian music-hall singer named Marie Ambrose. According to Andy's carefully crafted script, Marie had lived close to the prison, and her ghost had often been seen in the building. Andy then passed various objects around the group that were allegedly associated with her life, including a maraca, a handbell and a wicker ball. In reality, I had bought the objects from a local junk shop a few days before the shows. All of the objects, and the table around which everyone was seated, had small spots of luminous paint on them so that the group would be able to see them in the dark. Andy placed the objects on the table, asked everyone to join hands, and extinguished the candles. The room was plunged into complete darkness, but the objects on the table became visible from their slightly luminous glow. Andy slowly started to summon the non-existent spirit of Marie Ambrose.

The group was first asked to concentrate on the wicker ball. After a few minutes, it rose a few feet into the air, moved around the seance room, and gently returned to the table. Next, they turned their attention to the maraca, which, on a good night, slowly rolled across the table. These apparently ghostly phenomena were the result of the types of simple trickery that had been used by fake mediums at the turn of the twentieth century. It soon became obvious to us that they were still capable of having an impact on a modern-day audience. We filmed many of the seances with an infra-red camera, and the tapes showed that some people around the table gasped, some screamed, and many sat shaking in stony silence.

Then came the most important part of the evening. The suggestion. Andy asked Marie to make her presence known by moving the large, heavy table. The table remained completely stationary, but Andy suggested that it was levitating, using comments such as 'That's good, Marie', 'Lift the table higher', 'The table is moving now'. Andy then released the non-existent spirit of Marie back into the ether, the lights were turned on, and everyone was thanked for coming to the show.

Two weeks later, our guinea pigs were sent a questionnaire about their experiences during the show. We first asked people whether they thought that any of the events they had witnessed were actually paranormal. Forty per cent of people who expressed a prior belief in the paranormal thought that the phenomena were the result of genuine ghostly activity, compared with only about 3 per cent of disbelievers. We then examined whether the suggestion had been effective. The results were startling. Over a third of people described how they had actually seen the table levitate. Again, participants' prior belief or disbelief in the paranormal played a key role, with half of disbelievers correctly stating that the table didn't move, versus just a third of believers. Our questionnaire also asked people whether they had had any unusual experiences during the seance. It seemed that the atmosphere we had created caused people to experience a whole range of spooky effects, with one in five reporting cold shivers, a strong sense of energy flowing through them, and a mysterious presence in the room.

The message was clear. In the same way that simple suggestion can be used to fool people into recalling various illusory events during their childhood, so it could also make a significant proportion of people experience the impossible.

A few years after the seance show I teamed up with a

television company to explore whether the same sort of techniques could be used to create a belief in New Age mumbo-jumbo, and even get people to part with their hard-earned cash.

Before the study started, we visited a local hardware shop and bought two objects – a brass curtain ring worth fifty pence and a chrome light-pull for two pounds. The manager of a large shopping centre in Hertfordshire kindly allowed us to carry out the study in the middle of his mall.

This initial phase of the experiment was designed to establish a baseline. We stopped people, asked them to place the brass ring or light-pull in their hand, and tell us if they felt anything odd. Perhaps not surprisingly, no one reported a thing. It was time to introduce some suggestion.

I explained to the next set of passers-by that I was a psychologist, that I had designed two objects to make people feel slightly unusual, and was looking to road-test the designs. Again, people kindly placed them in their hands. This time the reaction was quite different. Whereas before we had encountered nothing but blank faces, now the suggestion began to play with their minds. People started to report all sorts of slightly odd effects. Some said that the objects made them feel relaxed. Others said that they caused a slight tingling sensation. Often they would get an effect with one object and not the other, and were keen to know the difference between the two. When I asked how much they would be prepared to pay for the objects, people estimated between five and eight pounds.

So far we had employed only verbal suggestion. Now it was time to add some visual elements into the mix. I donned a white laboratory coat, and bought two cheap boxes for the curtain ring and light-pull. I approached a variety of shoppers, and, once again, people were kind enough to help out. I explained that I was looking for people to provide honest feedback on the two devices, which were designed to elicit some strange feelings. This time the reactions were even more extreme. One person said that the brass curtain ring made him feel high. Another said that the chrome light-pull made him feel as if his hands were magnetic and attracted towards one another. Another said that she felt as if there were electricity running through her hands. It was a dramatic demonstration of how easily suggestion can be used to part the gullible from their cash. How much were people now prepared to pay for the fifty-pence brass ring or two-pound light-pull?

The estimates ranged between fifteen and twenty-five pounds.

3

Believing six impossible things before breakfast: Psychology enters the twilight zone

How superstition has cost millions and killed thousands, why seemingly improbable coincidences are surprisingly likely, how people really walk on red-hot coals, and the truth behind haunted houses and the dreaded 'brown note of death'.

The Savoy Hotel in London is famous for fine dining, attentive service, grandiose interiors, and, of course, a three-foot-high wooden black cat called Kaspar. In 1898, a British businessman named Woolf Joel booked a table for fourteen at the hotel. Unfortunately, one of his guests cancelled at the last moment, leaving him with just thirteen diners. Woolf decided to ignore the old wives' tale that it is unlucky to have thirteen people around a table, and pressed ahead with the meal. Three weeks later, he travelled to South Africa and was shot dead in a highly publicized murder. For decades after the incident, the Savoy didn't allow parties of thirteen to dine at the hotel, and went so far as to have a member of staff join any such group, rather than run the risk of having another murder on their hands. In the 1920s, the hotel asked designer Basil Lonides to produce a sculpture to replace their human good luck charm, and he created Kaspar. Since then, this beautiful Art Deco cat has been joining wealthy parties of thirteen for dinner. Each time, he is fitted with a napkin, given a full place setting, and served the same food as his tablemates. Apparently, he was a firm favourite with Winston Churchill, who helped secure his return when he was kidnapped during the Second World War by a group of high-spirited officers dining at the hotel.

Superstitious and magical thinking pervade our entire lives. Perhaps not surprisingly, the topic has attracted more than its fair share of strange and unusual research. The work has involved extensive interviewing of estate agents, observing fishermen in remote regions of New Guinea, playing 'pass the parcel' across an entire country, secretly introducing low-frequency sound waves into classical music concerts, and having a group of people attempt to walk across 60 feet of red-hot coals. The results have revealed why much of society believes in the impossible, why strange coincidences are actually surprisingly likely, and why people experience ghostly goings-on in allegedly haunted buildings.

Superstitious minds

Dr Samuel Johnson always tried to court good fortune by leaving his house right foot first, and avoided treading on cracks in the pavement. Adolf Hitler believed in the magical powers of the number seven. President Woodrow Wilson believed that the number thirteen had consistently brought luck into his life, noting that there were thirteen letters in his name, and during his thirteenth year at Princeton University he became their thirteenth president.[1] His Royal Highness, Prince Philip, apparently taps on his polo helmet seven times before a game. Swiss tennis ace Martina Hingis allegedly avoids stepping on the court 'tramlines' between points. American basketball star Chuck Persons admitted to feeling nervous before a game unless he had eaten two KitKats, or two Snickers bars, or one KitKat and one Snickers bar.[2] Even Nobel prize-winning physicist Niels Bohr is rumoured to have placed a horseshoe over his door. (Although here the evidence is debatable. When asked whether he thought it really brought him good luck, Bohr replied, 'No, but I am told it works whether you believe in it or not.')

Irrationality is not restricted to princes, politicians, and physicists. One recent Gallup Poll revealed that 53 per cent of Americans said that they were at least a little superstitious, and an additional 25 per cent admitted to being somewhat or very superstitious.[3] Another survey revealed that 72 per cent of the public said that they possessed at least one good luck charm.[4] The results of my own 2003 superstition survey, conducted in collaboration with the British Association for the Advancement of Science, revealed the same high levels of belief in modern-day Britain, with approximately 80 per cent of people routinely touching wood, 64 per cent crossing their fingers and 49 per cent avoiding walking under ladders.[5] Even some of America's brightest students engage in behaviour like this. Harvard undergraduates routinely touch the foot of the statue of John Harvard for good luck before going into their exams, whilst those at the Massa-chusetts Institute of Technology rub the nose of a brass image of inventor George Eastman. Over the years, both Harvard's foot and Eastman's nose have developed considerable superstition-induced shines.

Although the consequences of many traditional beliefs,

such as touching wood or carrying a lucky charm, are relatively harmless, the effects of other superstitious ideas have far more serious implications.

In early 1993, researchers wanted to discover whether it really was unlucky to live in a house numbered 13.[6] They placed advertisements in over thirty local newspapers, asking people living in 'Number 13' houses to get in touch, and rate whether their good fortune had decreased after moving to the house. Five hundred replied, with approximately one in ten reporting that they had experienced more bad luck as a result of moving to the unlucky number. The researchers then wondered whether the belief might affect house prices, and so conducted a national survey of estate agents about the issue. A surprising 40 per cent said that buyers were often resistant to buying property numbered 13, and that this often resulted in sellers having to lower the price of the properties.

At other times the effects can be a matter of life or death. In chapter 1 we met sociologist David Phillips, a scientist fascinated with investigating whether people's date of birth influenced their moment of death. In an article published in the *British Medical Journal*, Phillips reported a link between superstition and the precise moment of passing away.[7] In Mandarin, Cantonese, and Japanese, the words for 'death' and 'four' are pronounced in almost exactly the same way. Because of this, the number four is seen as unlucky in Chinese and Japanese cultures. Many Chinese hospitals do not have a fourth floor, and some Japanese people are nervous about travelling on the fourth day of the month. The link also stretches to California, where new businesses are offered a choice of the last four digits in their telephone numbers. Phillips noticed that Chinese and Japanese restaurants contain about a third fewer '4's than expected, a pattern absent in restaurants describing themselves as American. All of this led Phillips to wonder whether the superstitious stress induced on the fourth of each month might play an important role in health. Could it be linked, for example, to the onset of a heart attack?

To assess the possible effects of these beliefs on health, Phillips and his team analysed the records of over forty-seven million people who had died in America between 1973 and 1998. They compared the day of death of Chinese and Japanese Americans with white Americans, discovering that in the Chinese and Japanese populations, cardiac deaths were 7 per cent higher on the fourth of each month than any other day. This figure jumped to 13 per cent when the investigators focused on chronic heart deaths. The mortality data from white Americans contained no peaks. The work is controversial, and has been questioned by other researchers.[8] Nevertheless, Phillips and his team are confident that something strange is happening, and named the alleged effect after Charles Baskerville, a character in the Arthur Conan Doyle story *The Hound of the Baskervilles*, who suffers a fatal heart attack from extreme psychological stress.

It is one thing for superstitious people to inadvertently kill themselves, but quite another when their beliefs directly affect other people's lives. Thomas Scanlon and colleagues looked at traffic flow, shopping centres, and emergency hospital admissions on Friday 13ths. Over a two-year period, they discovered a significantly lower traffic flow on sections of London orbital M25 motorway on Friday 13th compared to Friday 6th, suggesting that nervous drivers might be staying indoors.[9] They then examined various types of hospital admission on the two dates, including poisoning, injuries caused by venomous animals, self-harm, and transport-related accidents. Of these, only the traffic accident grouping showed a significant effect, with more accidents on Friday 13th than Friday 6th. The effect is far from trivial, with an increase of 52 per cent on the fateful day. Scanlon and his colleagues had access to admissions from only one hospital, however, so the numbers were relatively small, and thus it was possible that their findings were simply down to chance. In a significantly larger but equally controversial study, Finnish researcher Simo Näyhä examined similar records between 1971 and 1997 for the whole of Finland.[10] During this time there were 324 Friday 13ths and 1339 'control' Fridays. The results supported the previous research, especially for women. Of the deaths for men, only 5 per cent could be attributed to the unlucky day, while for women the figure was a staggering 38 per cent. Both sets of researchers attribute the rise in accident rates to drivers feeling especially nervous on the most inauspicious of unlucky days. The message is clear: superstition kills.

The year of the Fire-Horse

Superstitious beliefs can also exert a significant effect on entire societies. According to the ancient Sino-Japanese almanac, each year is labelled on the basis of two elements: one of twelve animals (such as a sheep, monkey, or chicken) and ten heavenly stems (such as earth, metal, or water). The year of the Fire-Horse occurs just once every sixty years, which is perhaps just as well, because it symbolizes little but bad fortune. According to legend, any women born in this year will have a fiery temperament, making them highly undesirable wives. Although this notion stretches back into the mists of time, it is kept alive in modern-day Japan through a popular Kabuki drama based around the story of Yaoya-Oshichi. According to the story, in 1682 Oshichi fell in love with a young priest, and thought it best to start a small fire to help cement their love. Unfortunately, she was born in the year of the Fire-Horse, and the fire spread out of control and eventually destroyed almost the whole of Tokyo.

The last year of the Fire-Horse was in 1966, and Japanese researcher Kanae Kaku decided to use the opportunity to examine whether superstitious thinking had an impact on the entire population of Japan.[11] The answer was a resounding, and astounding, yes. The year 1966 saw a 25 per cent decrease in the Japanese birth rate (corresponding to almost half a million fewer babies born during the year), and an increase of more than 20,000 induced abortions. Subsequently, Kaku discovered that the effect was not confined to Japan, uncovering similar drops in the 1966 birth rates for Japanese people living in California and Hawaii.[12] Curious, Kaku dug deeper into the data, and discovered something even more remarkable.[13] According to legend, females born during the year of the Fire-Horse will lead especially unlucky and ill-fated lives. In 1966, there was no easy method for determining the sex of a child prior to birth, and thus the only way to ensure a dearth of female offspring would involve infanticide. Would parents actually be prepared to kill their female babies simply because of an age-old superstitious belief? Kaku examined the neonatal mortality rates from accidents, poisoning, and an external cause of violence between 1961 and 1967. The results were chilling. In 1966, the mortality rates for newborn girls, but not boys, were significantly higher than the surrounding years. These patterns caused Kaku to conclude that Japanese girls were indeed being 'sacrificed to a folk superstition' during the year of the Fire-Horse.

Japanese researcher Kenji Hira and his team from Kyoto University assessed the financial costs of another type of Japanese superstition.[14] Before 1873, Japan utilized a six-day lunar calendar, with each day designated as Sensho, Tomo-biki, Senpu, Butsumetsu, Taian, and Shakku. Even today, Taian is traditionally seen as a lucky day and Butsumetsu as an unlucky day. Because of this, many hospital patients wish to be discharged on Taian. Figures from three years of recent hospital admissions revealed that many patients were indeed extending their stay to ensure this outcome. The researchers estimated that this superstitious behaviour cost Japan approximately fourteen million pounds each year. And it is not only Japan. In Ireland there is a superstitious belief that if you leave a place on a Saturday, you are unlikely to be away for long ('Saturday flit, short sit'). An analysis of 77,000 Irish maternity records over four years revealed that about 35 per cent fewer patients than expected were discharged on Saturdays, whilst increases of 23 per cent and 17 per cent in discharges were observed on Fridays and Sundays respectively.[15]

The message is clear. Superstitious beliefs are not just about the harmless touching of wood or crossing of fingers. Instead, beliefs can affect house prices, the number of people injured and killed in traffic accidents, abortion rates, and monthly death statistics, and can even force hospitals to waste significant amounts of funding on unnecessary patient care.

Given the important implications of superstition, perhaps it isn't surprising that many researchers have examined just why it is that so many people allow irrational ideas to affect the way they think and behave.

Lotteries, lunacy, and the Thirteen Club

Proponents of superstition argue that there must be something to these beliefs because they have survived the test of time. They have a point. Lucky charms, amulets and talismans have been found in virtually all civilizations throughout recorded history. Touching wood dates back to pagan rituals that were designed to elicit the help of benign and powerful tree gods. When a ladder is propped up against a wall it forms a natural triangle that was seen as symbolic of the Holy Trinity, and to walk under the ladder was seen as breaking the Trinity. The number thirteen is seen as unlucky because there were thirteen people at Christ's Last Supper.[16]

Sceptics view this type of historical data, not as evidence of the validity of superstition, but rather of a depressingly deep-seated irrationality, noting that scientific tests of superstition have consistently obtained negative findings. They, too, have a point. The alleged relationship between superstitious behaviour and national lotteries is a good example. Each week, millions of people across the world buy lottery tickets in the hope of changing their lives for the better by getting lucky and winning a large sum of money. The winning numbers are drawn at random, and so there should be no way of predicting the outcome of the lottery. That, however, doesn't stop people trying all sorts of magical rituals to increase their chances of success. Some people always choose the same 'lucky' numbers each week. Others base their choice on significant events, such as their birthday or the ages of their children, or on their house number. A few have developed even more obscure rituals, including writing each of the numbers on pieces of paper, spreading them across the floor, letting their cat into the room, and choosing those touched by the cat.

When the National Lottery was first launched in Britain, I worked with fellow psychologists Peter Harris and Matthew Smith to put these various rituals to the test.[17] In a large-scale experiment conducted with a BBC television programme entitled *Out Of This World*, we asked 1,000 lottery players to send us their numbers prior to a draw, to indicate whether they thought themselves lucky or unlucky, and to describe the method they had used make their selection. The lottery forms were returned remarkably quickly. In all, we received replies from 700 people who, between them, intended to buy just over 2,000 lottery tickets. Matthew and I entered everyone's choice of numbers into a giant spreadsheet one day before the draw. Having done so, we suddenly realized that we had collected some extraordinary information. If lucky people really do pick more winning lottery numbers than unlucky people, then the numbers that were being chosen by the lucky people, but not by the unlucky ones, would be more likely to be winning numbers. It hadn't occurred to us before, but if the theory was right, some of the data we had collected for our experiment could make us millionaires.

Matthew and I debated the ethics of the situation for at

least a few seconds, and then started to analyse the information. We noticed that some numbers were being chosen by lucky people and avoided by unlucky people. We slowly identified the 'most likely' winning numbers – 1, 7, 17, 29, 37, and 44. For the first and only time in my life, I bought a lottery ticket. The UK National Lottery draw takes place every Saturday night and is broadcast live on primetime television. As usual, the forty-nine balls were placed in a rotating drum and six balls, plus a special 'bonus' ball, were randomly selected. The winning numbers were: 2, 13, 19, 21, 45, 32. We hadn't managed to match a single number. But had the lucky and unlucky people in our experiment fared any better? Actually, there was no difference. Lucky people did no better than unlucky people, and those using any kind of superstitious ritual were just as unsuccessful as those choosing their numbers randomly. There were also no differences between people basing their decisions on their date of birth, the ages of their children, or the behaviour of their pets. In short: Rationality 1 – Superstition 0.

Other researchers have taken more unusual approaches to the issue. One of my favourite experiments was conducted by an American high school student named Mark Levin.[18] Levin and his friends set out to test the popular belief that a black cat crossing your path brings bad fortune. First they asked people to measure their fortune by playing a simple computerized coin-tossing game in which they had to guess whether a coin would land heads or tails. Next, highly skilled cat wranglers ensured that a black cat walked across these people's paths as they ambled along a corridor. Finally, all of the participants played the coin-tossing game a second time in order to reassess their luck. After much coin-tossing and cat-crossing, the results revealed that the black cat had absolutely no effect. To make absolutely certain that they had left no stone unturned, the researchers repeated the experiment with a white cat, and again obtained null results. Levin ended his article by noting that critics of his experiment might argue that the bad fortune associated with a black cat manifests itself only in real-life situations, and not in an experiment involving a coin-tossing game, but refutes the idea, noting: 'I own a black cat, and although she has crossed my path hundreds of times, I see no degradation in my school work or social life.'

Similar work has been carried out in those most apparently rational of places – hospitals. Medics are a surprisingly superstitious lot, as is shown by work on alleged behavioural effects associated with a full moon. A team of American researchers examined just under 1,500 records of trauma victims admitted to hospital throughout a full year, and discovered no relationship between the moon being full and the number of admissions, mortality rate, type of injury, or length of stay.[19] Despite this, a survey carried out in 1987, and reported in a paper simply entitled *Lunacy*, revealed that 64 per cent of emergency physicians were convinced that the full moon affected patient behaviour.[20] Of these, 92 per cent of nurses reported finding working during a full moon more stressful, although there is reason for scepticism about this latter finding, given that the same people also argued that such stress justified 'lunar pay differentials'.

It doesn't stop there. In the same way that well-wishing is considered unlucky in the theatre (causing actors to tell their fellow thespians to 'break a leg'), so doctors working on emergency wards consider that comments such as 'Looks like it will be a quiet night' can cause a flood of new patients. This superstition was put to the test by Andrew Ahn from the Massachusetts General Hospital and his colleagues, and described in the pages of the *American Journal of Medicine*.[21] Thirty medics were randomly assigned to one of two groups. A 'jinxed' group received a message stating 'You will have a great call day', whilst those in the control group received a blank piece of paper.

The jinxed group did not receive any more admissions, or any less sleep, than the control group (if anything, those receiving the message seemed to have fewer patients and more sleep than those receiving the blank piece of paper). As with all important findings in science, this work has now been replicated in different parts of the world. In one test, British medics Patrick Davis and Adam Fox randomly assigned each of their days on an emergency ward to one of two conditions: a control day or a 'Q' day.[22] During the control days, the team discussed the weather, whilst during a 'Q' day they spoke about how they all thought that it would be a quiet night. In line with the results obtained by their American counterparts, the British medics found no significant differences in the numbers of admissions between the two conditions.

Perhaps the most systematic and thorough test of superstition dates back to the turn of the last century. In the 1880s, American Civil War veteran Captain William Fowler decided to tempt fate by creating a Thirteen Club in New York.[23] The idea was simple. He would invite twelve guests to join him for dinner on the thirteenth day of each month, and break various widely held superstitions such as spilling salt on the table, crossing forks, and opening umbrellas indoors. The scheme was an instant success and quickly became one of

New York's most popular social clubs, forcing Fowler to hire ever-larger rooms capable of holding several tables each containing thirteen guests. Over the next forty years or so, the club's membership ran into the thousands, and its list of honorary members included no less than five successive American presidents. The members' strength of feeling against the superstitious mindset should not be underestimated. In a speech to the club on 13 December 1886, politician, agnostic, and orator Robert Green Ingersoll noted:

> The most important thing in this world is the destruction of superstition. Superstition interferes with the happiness of mankind. Superstition is a terrible serpent, reaching in frightful coils from heaven to earth and thrusting its poisoned fangs into the hearts of men. While I live, I am going to do what little I can for the destruction of this monster.[24]

Ingersoll went on to explain that if he died, and discovered that there was an afterlife, he would spend his time there continuing to argue with those that believed in the supernatural. Despite consistently engaging in behaviours that allegedly attracted little but ill fortune, death and disease, the members of the Thirteen Club proved remarkably healthy and happy. At the club's thirteenth dinner in 1895, Fowler reported that the death rate of club members was slightly *below* that of the general population. The positive effects of breaking superstitious taboos were underlined by the comments made by one-time club leader J. Arthur Lehman in 1936:

> My advice to anyone that wants real luck and happiness and health is to break every possible known superstition today . . . All of the members of the Club that I can

remember had good luck . . . I'm 78 now and I defy you to find anyone happier or healthier than I am.

So if superstitions have no validity, why have they survived the test of time, and been passed down from generation to generation? Part of the answer takes us from the islanders of the coast of New Guinea, to Israelis trying to cope with Scud attacks during the first Gulf War.

Melanesians and missiles

Bronislaw Malinowski was one of the world's greatest anthropologists. He grew up in Poland and originally studied mathematics and the physical sciences. A chance encounter, however, changed the course of Malinowski's life. Whilst preparing for a foreign language exam, he came across a copy of *The Golden Bough* by respected anthropologist Sir James Frazer. Frazer's book was a detailed study of magic and religion in diverse cultures across the world. The book persuaded Malinowski to travel to Britain to pursue a career in anthropology. Partly to escape possible internment at the outbreak of the First World War, Malinowski travelled to Melanesia, a small island off the coast of New Guinea, to immerse himself in the culture of an isolated community known as the Trobriand Islanders. His book describing his subsequent work there, *Argonauts of the Western Pacific*, is now universally regarded as a masterpiece.[25] Malinowski studied many aspects of the Trobrianders' daily lives, and was especially interested in one aspect – their superstitious behaviour. He noticed that the Trobrianders used standard fishing techniques when they worked in the relatively calm waters of a lagoon, and engaged in elaborate magical and superstitious rituals only when they ventured onto the much more dangerous open seas. Malinowski speculated that superstitious behaviour had its roots in the unpredictability of their lives. When fishing on the lagoons, the Trobrianders faced little uncertainty. They felt in control of the situation and so did not feel the need to engage in any superstitious behaviour. However, the situation was completely different on the open seas. Here they knew that life was far more unpredictable, and so attempted a variety of magical rituals in a vain attempt to get control of the situation and decrease the danger. In short, Malinowski believed that superstition comforted the islanders by providing them with a sense of control over the slings and arrows of outrageous fortune.

It might be nice to believe that irrationality was confined to a group of isolated islanders in the 1920s, but exactly the same pressures that forced the Trobrianders to carry out elaborate rituals in the open seas around Melanesia also cause us to touch wood, cross our fingers, and reach for the lucky rabbits' feet.

By the middle of the 1920s, inflation in Germany was so high that paper money was carried in shopping bags, and people were eager to spend any money the moment that they had it, for fear that it would be severely devalued the following day. By 1932, almost half of the population were unemployed. In 1982, Vernon Padgett from Marshall University and Dale Jorgenson from the California State University published a paper comparing the number of articles on astrology, mysticism, and cults appearing in major German magazines and newspapers between the two world wars, and the degree of economic threat each year.[26] Articles on gardening and cooking were also counted as controls. An index of economic threat was calculated on the basis of wages, percentage of unemployed trade union members, and industrial production. When people were suffering an economic downturn, the number of articles on superstition increased. When things were going better, they decreased. The strong relationship between the two factors caused the authors to conclude that

> . . . just as the Trobriand islanders surrounded their more dangerous deep sea fishing with superstitions, Germans in the 1920s and 1930s became more superstitious during times of economic threat.

The authors link their findings with much broader social issues, noting that in times of increased uncertainty, people look for a sense of certainty and this need can cause them to support strong leadership regimes, and believe in various irrational determinants of their fate, such as superstition and mysticism.

A study carried out in Israel by psychologists at the University of Tel Aviv during the 1991 Gulf War graphically illustrated the same idea.[27] Soon after the start of the war it became clear that some cities, like Tel Aviv and Ramat Gan, were in danger of attack from Scud missiles, whilst others, like Jerusalem and Tiberias, were relatively safe. The researchers wondered whether the increased stress associated with living in the more dangerous areas would encourage people to become more superstitious. To test this idea they put together a questionnaire about superstition. Some of the questions asked about well-known examples of magical thinking, such as whether it was a good idea to shake hands with a lucky person, or to carry a lucky charm. Others concerned new forms of superstitious behaviour that had emerged since the start of the attacks. For example, since the mid 1980s, buildings in Israel had been constructed with a room capable of being sealed with plastic in order to protect its occupants from a gas attack. The questionnaire asked people whether they thought it best always to step into the sealed room right foot first, and whether the chances of being hit during an attack were greater if a person whose house had already been hit were present in the sealed room. Next, the researchers went door to door in both high-risk and low-risk areas asking about 200 people whether they carried out these behaviours. The researchers' speculations were confirmed: people living in areas subjected to severe missile attacks had developed far more superstitious beliefs and behaviours than those living in less dangerous parts of the country.

The research from New Guinea, Germany and Israel all suggests that many people become superstitious to help them cope with uncertainty. However, other work shows that these beliefs can also develop for quite different reasons, and have far more negative consequences.

Contagious thinking

Sir James Frazer, in his classic text on magic and religion, describes several types of magical thinking. One of the most fundamental is the 'law of contagion'. According to the theory, once an object has been in contact with a person, the object somehow possesses the 'essence' of that person.

In certain forms of magical ritual, a person trying to cast a spell may try to obtain the hair, or fingernail clippings, of an intended victim, and use this to exert some kind of (usually negative) influence over them.

Psychologist Paul Rozin and his colleagues at the University of Pennsylvania wondered whether this way of thinking was alive and well in modern-day Western society, and whether it might even underlie certain types of prejudice and irrationality.[28] To find out, they conducted a series of unusual, but insightful, experiments. The researchers asked people to

> Rate how you would feel about wearing a nice, soft, blue sweater, big and bulky, unisex in style. It was laundered a couple of days ago, but it's new, has never been owned or worn by anyone.

Not surprisingly, people said that they had no problem wearing the sweater. The experimenters then asked them to imagine that the sweater had been worn by someone who had contracted AIDS through a blood transfusion. The experimenters said that the sweater had been laundered a couple of days ago, and that the person with AIDS had worn it for only thirty minutes, but suddenly people really didn't want to wear the sweater. Even though they knew there was no health or hygiene issue, the superstitious theory of contagion kicked in, and they could not bring themselves to wear it. Rozin and his colleagues varied the imaginary sweater owners, and discovered that the idea of the sweater having once belonged to someone who personified evil, such as a mass murderer or fanatical leader, elicited the strongest reaction from people. In fact, Rozin's results revealed that people would rather wear a sweater that had been dropped in dog faeces and not washed (raising genuine health concerns), than a laundered sweater that had once belonged to a mass murderer.

It's a small, small world, and year-on-year shrinkage

People often develop magical beliefs about the world because they have experienced something seemingly weird. With the concept of coincidences, events appear to coincide in a way that both seems meaningful, and defies the odds. One of the best-known sets of coincidences surrounds the deaths of American Presidents John Kennedy and Abraham Lincoln. Lincoln was killed in the Ford Theatre, whilst Kennedy was assassinated when he was travelling in a Lincoln car, built by Ford. Lincoln was elected to Congress in 1846, Kennedy in 1946. Lincoln was elected president in 1860, Kennedy in 1960. The surnames of both men contain seven letters, and both killings took place on a Friday. After their deaths, both presidents were succeeded by men named Johnson. Andrew Johnson was born in 1809, Lyndon Johnson in 1909.

These sorts of amazing moments do not restrict themselves to American presidents, but instead pop up from time to time in most people's lives. In the 1920s, three strangers were travelling through Peru by train. Sitting in the same carriage, they introduced themselves to one another, only to find that the first man's surname was Bingham, the second man's was Powell, and the third man's was Bingham-Powell. Another remarkable coincidence took place at London's Savoy Hotel, home of Kaspar the lucky black cat, in 1953. Television reporter Irv Kupcinet was staying at the hotel to cover Elizabeth II's coronation. Opening one of the drawers in his room, he found some items belonging to his friend Harry Hannin, manager of the well-known basketball team, the Harlem Globetrotters. Just two days later, Kupcinet received a letter from Hannin in which he explained that he had been staying at the Hotel Meurice in Paris, and had found one of Kupcinet's ties in a drawer in his room. Faced with such curious incidents, many people would simply ask, 'What are the chances of that?', and leave it there. But some scholars, such as Stanford mathematician Persi Diaconis, have delved deeper.

Diaconis has been invited by Vegas casinos to determine whether their card-shuffling machines really do randomize the order of decks (they didn't), used high-speed cameras taking 10,000 frames per second to analyse human coin-tossing (revealing that coins show a tiny bias towards landing the same way that they started), and persuaded a team of Harvard technicians to create a machine capable of producing a perfectly random coin-toss. He has also written one of the seminal papers on the mathematics and psychology of coincidences, arguing that certain little-known statistical laws make some seemingly impossible events surprisingly likely. The law of large numbers is one example.

Almost every week in Britain a truly amazing coincidence takes place – an event that we know is extremely unlikely to happen by chance alone. In fact, the odds of this event happening are a staggering fifteen million to one. Someone wins the lottery jackpot. Why does this unlikely event routinely happen week after week? Because a huge number of people buy lottery tickets. It is exactly the same with many coincidences. There are millions of people in the world living complex lives, and so it is not surprising that once in a while someone wins the coincidence jackpot and experiences a genuinely unlikely event. Although it is tempting to see these events as a sign from the gods, or as evidence of a mysterious sense of connection between people, in reality it may all be down to chance. Arthur Conan Doyle put it beautifully in *The Adventure of the Blue Carbuncle*:

> Amid the action and reaction of so dense a swarm of humanity, every possible combination of events may be expected to take place, and many a little problem will be presented which may be striking and bizarre.

The same precept also applies to amazing anagrams that seem to contain hidden messages, or wonderfully succinct descriptions of people or events. The words 'US president Ronald Reagan' are a precise anagram of 'repulsed and ignorant arse', whilst 'President Clinton of the USA' can be scrambled to make 'to copulate he finds interns'. My favourite anagram was discovered by puzzle creator Cory Calhoun, and involves the famous phrase from Shakespeare's *Hamlet*: 'To be, or not to be: that is the question:/ Whether 'tis nobler in the mind to suffer/ The slings and arrows of outrageous fortune'. This is an exact anagram of a statement that provides a perfect summary of the entire play: 'In one of the Bard's best-thought-of tragedies, our insistent hero, Hamlet, queries on two fronts about how life turns rotten.' Although these examples may look amazing, there is nothing magical taking place. It is simply the law of large numbers at work. Given the large number of combinations of letters in the words, and the huge amount of text in plays and books, it is not surprising that once in a while amazing anagrams emerge. What *is* perhaps more surprising is that some people are prepared to invest significant amounts of their time looking for them.

Although the law of large numbers accounts for many coincidences, sometimes there is a deeper psychology at work. A 1993 survey showed that one of the most frequently experienced coincidences is that of the 'small world' phenomenon, in which strangers meet at a party, only to discover that the two of them have a mutual acquaintance.[29] Almost 70 per cent of people claimed to have had this experience, with about 20 per cent experiencing it frequently. In the 1960s, the phenomenon intrigued the well-known American psychologist, Stanley Milgram.

Milgram was a remarkable man and responsible for conducting some of the world's most famous psychology experiments. Starting in late 1960, Milgram carried out a series of studies examining whether ordinary people would be prepared to inflict pain and suffering on others, simply because they were told to do so by an experimenter.[30] In the study, an experimenter asked participants to deliver increasingly dangerous electric shocks to another participant (actually an actor who was simply pretending to receive the shocks). If the participant expressed any concern about what they were doing, the experimenter encouraged them to persevere with the procedure, using phrases such as 'Please continue' and 'The experiment requires you to go on'. Milgram's results revealed that about 60 per cent of participants were prepared to deliver what they thought was a potentially lethal shock to their hapless victim, because they were told to do so by a man wearing a white coat. Milgram's electric shock study is very well known. It can be found in almost every introductory psychology textbook, and is one of the very few behavioural studies to have exerted a significant effect on popular culture. In the mid 1970s, CBS broadcast a dramatization of the electric shock experiments with William Shatner playing the role of Milgram, and in 1986, musician Peter Gabriel wrote a song entitled 'We do what we're told (Milgram's 37)', referring to one of Milgram's experiments in which thirty-seven out of forty participants were fully obedient. What is not so well known is that his work inspired several equally striking follow-up studies. Professors Sheridan and King were concerned that participants may have correctly guessed that the participant receiving the shocks was actually an actor, and so repeated the study in the 1970s using genuine shocks being administered to real puppies.[31] The resulting paper, entitled *Obedience to Authority with an Authentic Victim*, describes how just over 50 per cent of men delivered the maximum shock to the puppies, versus 100 per cent of women.

Milgram continued to devise and carry out unusual and thought-provoking experiments throughout his academic career. In fact, he had developed such a reputation for such work that when he burst into a colleague's lecture theatre on 22 November 1963, announcing the assassination of Kennedy, many of the students assumed that it was all part of yet another Milgram experiment.[32]

Building on theoretical work being undertaken at the Massachusetts Institute of Technology, Milgram decided on a hands-on approach to trying to understand what might lie behind the 'small world' phenomenon.[33] A letter was sent to 198 people living in Nebraska, who were asked to help ensure that it made its way to a 'target person' – a named stockbroker who worked in Boston, and lived in Sharon,

Massachusetts. There was, however, a catch. Participants could not mail the letter directly to the stockbroker; they were allowed to send it only to someone whom they knew on first-name terms and whom they thought might know him. Each recipient was asked to do the same, and, again, was allowed to send the package on only to someone that they knew on first-name terms.

How many people were needed to link complete strangers? Given the tens of millions of people in America, many were surprised to discover that there tended to be just six people linking the initial volunteer and the target person – giving rise to the popular notion that we are all connected by just six degrees of separation. The results suggest that society is much more closely knit than might at first be imagined, and helps explain why jokes, gossip, and fads can rapidly spread simply by word of mouth. In addition, by examining the relationship between the people in each of the completed chains, Milgram was able to gain some insights into the social structure of 1960s America. People were far more likely to pass the parcel on to someone of their own, rather than the opposite, sex, and most links involved friends and acquaintances rather than relations. The implications of Milgram's findings are not limited to social systems, but also explain a diverse range of other networks, including the operation of power grids, the spread of disease, the way in which information is passed around the Internet, and the neural circuitry underpinning brain functioning.[34]

Writing about Milgram's work in 1995, mathematician John Allen Paulos notes:

It's not clear how one would carry out studies to confirm

this, but I suspect that the average number of links connecting an arbitrary pair of people has shrunk over the last fifty years. Furthermore, this number will continue to shrink because of advances in communication and despite an increasing population.[35]

Given the importance of Milgram's giant game of pass-the-parcel, and Paulos's speculations about a shrinking world, it's surprising that hardly any researchers have attempted to repeat the experiment. So in 2003, my colleague Emma Greening and I teamed up with Roger Highfield, science editor of the *Daily Telegraph*, and the Cheltenham Science Festival, to address this issue.[36] We wanted to carry out the first British replication of Milgram's classic piece of quirkological research, and test two ideas. First, would we obtain the same number of links, or even, as Paulos suggested, fewer than Milgram? Secondly, was it possible to use the phenomenon to explain another mystery that had emerged during my study of lucky and unlucky people? Lucky people claim to have lots of chance encounters, and these seem to have a remarkably beneficial effect on their lives. They bump into someone at a party, discover that they know people in common, and, from these connections, end up getting married or doing business, for example. Or, when they need something, they always seem to know someone who knows someone who can solve their problem. In contrast, unlucky people rarely report such experiences. We wondered whether lucky people have lots of 'small world' experiences because they know lots of people and so are, without realizing it, making their own good fortune by constructing, and inhabiting, an especially small world.

I published a short article in the *Telegraph*, inviting readers who wished to participate in a 'small world' experiment to contact me. One hundred volunteers were then sent a package containing some instructions along with a set of postcards and envelopes. The instructions explained that the purpose of the experiment was to ensure that the parcel made its way to a certain 'target person'.

Rather than using a stockbroker in Boston, our target person was Katie Smith, a 27-year-old events organizer working in Cheltenham. As with Milgram's original study, all initial volunteers and subsequent recipients were asked to send the parcel only to someone they knew on first-name terms. The original participants, and all of their subsequent recipients, were also asked to return one of the postcards to us, so that we could track the packages as they moved around the country.

There tended to be just four people linking our initial volunteers to Katie – two less than obtained in Milgram's experiment. Some of the chains in our study provide striking illustrations of just how well-connected apparent strangers actually are. For example, one of our initial volunteers was a textile agent called Barry. Barry lives in Stockport and, perhaps not surprisingly, doesn't know Katie Smith. Barry passed the package on to his friend Pat because she lived close to Cheltenham racecourse. Pat also doesn't know Katie. She sent the package to her friend David, however, who happened to be head of the Cheltenham Science Festival. Bingo! David knew Katie and so was able to complete the chain and pass the package directly to her.

Our study was the first British replication of Milgram's famous experiment. The decrease in the average number of links taken to reach our target person may be owing to Britain being better connected than America. Alternatively, the results could be seen as supporting the intriguing possibility that the world has become substantially smaller over the last forty years. Perhaps, as a result of vast increases in electronic communication, telephone networks and travel, we are all connected to one another as never before. Maybe, on a social level, science and technology have genuinely shrunk the world.

Possible evidence of global shrinkage is all well and good, but did we find any evidence to suggest that lucky people are especially well connected, and therefore living in a smaller world than most? To find out, we asked each initial volunteer involved in the study to rate their general level of luckiness prior to taking part. Thirty-eight volunteers did not send their parcel to anyone, thereby guaranteeing that their packages would never reach Katie. Interestingly, the vast majority of these people had previously rated themselves as unlucky. We wanted to discover what lay behind this curious behaviour. These volunteers had gone to considerable lengths to ensure that they participated in the study, but had then effectively dropped out at the very first stage. We wrote asking why they had failed to send on the parcel. Their replies were telling – the majority said that they couldn't think of anyone they knew on first-name terms who could help deliver the parcel. As a result, from the outset it appears that the lucky participants knew far more potential recipients for the parcels than the unlucky people and were far more successful when it came to forwarding them. These results provide substantial support for the notion that lucky people are living in a much smaller world than unlucky people and that this, in turn, helps maximize their potential for 'lucky' small world encounters in life.

Walking on hot coals and things that go bump in the night

Some people appear to be able to walk on fire, crossing unharmed a long bed of burning coals with a surface temperature of approximately 1000°F. The scientific explanation for this amazing feat is that the thermal conductivity of coal is very low and, providing the bed of embers is relatively short, very little heat will be transferred to the walkers' feet. Many firewalkers, however, earn a good living expounding a more extraordinary explanation. According to them, they use the power of their mind to create a magical 'energetic' force field that protects them from harm, and claim to be able to teach this skill to others. Whereas science would predict that people could walk across approximately 15 feet of embers without being burnt, the paranormalists boast that they can walk any distance safely.

In 2000, I worked with the BBC science show, *Tomorrow's World*, to stage a dramatic test of this claim.[37] The programme spent a large sum of money burning 50 tons of wood to create a 60-foot-long bed of red-hot embers. Live on television, the alleged miracle-mongers put their paranormal theory to the test, which resulted in each of them jumping off the bed around the 25-foot mark with second-degree burns to their feet. I interviewed the firewalkers afterwards and discovered that they had their own explanations for their failure. One spoke about how the bright television lights had prevented him from entering into the deep trance needed for a successful demonstration. Another explained that her guardian angel had unexpectedly left her a few moments before the start of the walk. It was a remarkable demonstration of how belief in the impossible can be bad for your health, and even second-degree burns hadn't caused them to question their allegedly paranormal abilities.

Fortunately, most people do not think that they possess superhuman abilities. Many, however, believe that they have experienced equally strange phenomena. About a third of people believe in ghosts, and around one in ten claim to have actually encountered one. I have no idea whether ghosts actually exist, but I am fairly sure that people are quite capable of fooling themselves into believing that this is the case. For many years my colleagues and I have carried out various unusual experiments into the psychology of ghostly experiences.[38] Britain has more than its fair share of haunted homes, and much of the work has taken place at some of the best-known 'haunted' locations in the country. We were the first researchers to be invited to investigate the alleged ghostly goings-on at an official royal palace, spending ten days at the splendid Hampton Court on the outskirts of London. On another occasion, we staged a study in a series of apparently haunted vaults deep under the historic streets of Edinburgh, in Scotland.

People always seem a little disappointed to discover that our experiments are quite different to the sorts of studies portrayed in the film *Ghostbusters*. We tend not to wander around in jumpsuits with vacuum cleaners strapped to our backs, and, just for the record, have never caught a spirit form in a ghost trap. We don't set out, however, to prove or disprove the existence of ghosts. Instead, our work is all about trying to understand why people consistently report odd experiences in certain parts of these allegedly haunted locations.

Most of the studies have involved getting members of the public to walk carefully through the locations in a systematic way, and to describe any strange and unusual phenomena that they experience. Then, by examining the type of people who report these experiences, and the places in which they tend to report them, one can slowly start to piece together the psychology of the haunting.

We have discovered that some people are far more sensitive to the presence of alleged ghosts than others. Many volunteers will wander through a 'haunted' location and experience absolutely nothing, whilst a few moments later another person will walk through exactly the same spot, instantly feel uneasy and report a weird sense of presence. Those who experience strange phenomena tend to have very good imaginations. They are the type of people who make excellent hypnotic subjects, and often cannot remember whether, for instance, on leaving the house they have actually turned off their iron or simply imagined doing so. It seems that they are able to convince themselves that a spirit may really be standing right behind them, or hiding in a dark alcove. As a result, they feel genuinely scared and cause their bodies and brains to produce lots of the signals associated with fear, such as the hairs on the back of their neck standing up and a sudden sense of cold.

The work also suggests that context plays a vitally important role in the proceedings. This was brilliantly illustrated in an experiment published in 1997 by a collaborator of mine, American psychologist Jim Houran.[39] Jim took over a disused cinema that had absolutely no reputation for being haunted, and had two groups of people walk around it and rate the number of unusual phenomena they experienced. One group of people were told that the place was associated with lots of ghostly phenomena and so were on the lookout for non-existent spirit activity. The others were told that the theatre was currently undergoing renovation, and that they were there to rate how each room made them feel. Each of the two groups visited exactly the same locations in the cinema, but perceived them through completely different mindsets, causing the 'ghostbuster' group to report significantly more unusual experiences than the other group.

So does that mean that all ghostly experiences are the product of an overactive imagination combined with the correct context? Not necessarily. Other work, carried out by the late Vic Tandy, suggested that some ghostly experiences really may have been the result of something strange in the air.[40] Vic was an electrical engineer by training, and spent much of his time looking into phenomena that pricked his curiosity, including conjuring and ghosts. In 1998, he was working for a company that designed and manufactured life support equipment for hospitals. The firm ran a small laboratory that Vic shared with a couple of other scientists. This laboratory had a reputation for being haunted, with various cleaning staff reporting feeling rather odd in the building. Vic had always put this down to suggestion, or perhaps the effect of the various small furry animals that inhabited parts of the building. That was, until he himself had a rather strange experience. Working alone late at night, he started to feel increasingly uncomfortable, and cold. Next, he had the distinct impression of being watched, and looked up to see an indistinct grey figure slowly emerge in the left side of his peripheral vision. The hair on the back of his neck stood up, and, as he recalled, 'It would not be unreasonable to suggest that I was terrified.' Vic eventually built up the courage to turn and look at the figure. As he did, it faded away and disappeared.

Being the good scientist he was, Vic thought that maybe some of the bottles carrying anaesthetic agents might have leaked, causing him to hallucinate. A quick check revealed that this wasn't so. Stumped and stunned, he went home.

The following day, he was going to enter a fencing competition and so brought his foil into the lab for last-minute repairs. As he clamped the foil into a vice, it started to vibrate frantically. Although some might have been tempted to attribute the movement to poltergeist activity, Vic again searched for a rational explanation. This time, he found one. By carefully sliding the vice along the floor he was able to observe that the movement was at its maximum in the centre of the laboratory, and petered out towards each end of the room. Vic figured out that the room contained a low-frequency sound wave that fell below the human hearing threshold. Further investigation confirmed his suspicions. He traced the source of the wave back to a newly fitted fan in the air extraction system. When the fan was switched on, the fencing foil vibrated. When the fan was turned off, the foil remained stationary. But could Vic's discovery explain the seemingly ghostly phenomenon?

Vic knew that although these waves, usually referred to as 'infrasound', can't be heard, they carry a relatively large amount of energy, and so are capable of producing weird effects. In the 1960s, NASA were eager to discover how the infrasound produced by rocket engines might affect their astronauts during launching. Their tests showed that it did possess the potential to vibrate the chest, affect respiration, and cause gagging, headaches and coughing. Additional work suggested that certain frequencies can also cause vibration of the eyeballs, and therefore distortion of vision. The waves can move small objects and surfaces, and can even cause the strange flickering of a candle flame. Writing about his experiences in the pages of the *Journal of the Society for Psychical Research*, Vic speculated that some buildings may contain infrasound (perhaps caused by strong winds blowing across an open window, or the rumble of nearby traffic), and that the strange effects of such low-frequency waves might cause some people to believe that the place was haunted.

The idea is plausible, because infrasound is deeply strange. It can be produced naturally from ocean waves, earthquakes, tornadoes, and volcanoes. The 1883 Krakatoa eruption produced infrasound that circled the globe several times and was recorded on instruments worldwide. These low-frequency sound waves are also a by-product of nuclear explosions – hence the network of infrasonic listening posts that constantly monitor the environment for possible evidence of nuclear bomb tests.

Many animals are sensitive to frequencies undetectable by the human ear, including both ultrasound (high-frequency) and infrasound (low-frequency) sounds. Work on the detection, and use, of these extreme vibrations within the animal kingdom has a long history. In the early 1880s, Victorian scientist Francis Galton placed an ultrasonic whistle in the end of his hollow walking stick, and wandered around Regent's

Park Zoo noting down which animals responded to the high-frequency sounds produced whenever he pressed a rubber bulb at the top of the stick. After using this forerunner of the modern-day dog whistle, Galton reports that '. . . some curiosity is inevitably aroused by the unusual uproar my perambulations provoke in the canine community'. More recent, but conceptually similar, research has shown that whales, elephants, squid, guineafowl, and rhinoceros are all sensitive to low-frequency sounds, using these signals to migrate and to communicate over vast distances. This, combined with the fact that infrasound is a natural by-product of some earthquakes and tornadoes, led some researchers to question whether they might be able to detect the infrasound emitted from such natural disasters, and use it as a kind of early warning system. Some have suggested that this infrasound might account for the alleged fleeing of animals before the 2004 tsunami in Asia.

Low-frequency sound has also been investigated by the military as a possible basis for acoustic weaponry, and is informally referred to as the dreaded 'brown note' because it can allegedly vibrate people's bowels, causing them to defecate. Although sound engineers have known about this possibility for years, in 2000 the concept entered the public domain when an episode of the cartoon *South Park* centred on its child characters inadvertently broadcasting the note on American radio, causing the entire nation simultaneously to empty its bowels. Because of the resulting coverage, the American science show, *Myth Busters*, tested the concept by subjecting people to high levels of infrasound. Although the presenters reported feeling nauseous, the study failed to produce the much-rumoured effect.

There was, however, one problem. Most of the military and industrial work had used very high levels of infrasound, whereas Vic was speculating that much lower levels might be enough to induce a weird ghostly experience or two. It was time for an experiment.

A small draught, or a cost-effective way of finding God

Sarah Angliss, a long-standing friend of mine, studies acoustics, and produces sound installations for museums and other public spaces. One evening we were chatting about ghosts, and Vic's low-frequency sound hypothesis. Sarah was also interested in infrasound, and suggested that we team up and conduct an experiment. We needed an event that would attract large numbers of people, and one in which they could rate how they felt while infrasound was either present or absent. Sarah had the idea of piping infrasound into certain pieces being played at a live concert, to discover whether the secret sound wave affected the way the audience felt about the music. Could it, for example, induce the types of strange experiences often associated with the presence of a ghost, such as a sense of presence, sudden feelings of cold, and a tingling on the back of the neck?

Sarah led a crack squad of engineers and physicists to build a high-tech infrasound wave generator that allowed us to produce infrasound at will. In reality, this was a 7-metre-long sewage pipe with a low-frequency speaker in the middle. Sarah was present when the system was first turned on, noting:

. . . the pipe began to resonate strip lights, furniture and other loose odds and ends. As the pipe made very little audible noise, this was an odd experience. Seeing objects vibrate for no apparent reason, it is easy to imagine how infrasonic energy could be mistaken for a ghostly sighting.[41]

Teaming up with my PhD student at the time, Ciarán O'Keeffe, and National Physical Laboratory acousticians Dr Richard Lord and Dan Simmon, we hired one of the main concert rooms on London's South Bank and staged two unusual concerts.

The plan was simple. Each concert would consist of various pieces of contemporary piano music, performed by acclaimed Russian pianist, GéNIA. At four points during the concert, the audience would be asked to complete a questionnaire that measured their emotional response to the music, and note down any unusual experiences, such as a tingling sensation, or suddenly feeling cold. Just before two of these points, the auditorium would be flooded with infrasound. The two concerts would be identical, except for the timing of the infrasound. If the generator was turned on in one piece in the first show, it would be turned off in that piece in the second show. This counter-balancing procedure would enable us to minimize other sources of emotional effects, such as differences between the pieces of music. We would also be careful to produce a level of infrasound that was on the cusp of perception and this, coupled with the fact that it was masked by GéNIA's music, would help ensure that the audience were never consciously aware of its presence.

Staging the concert was far from easy. The South Bank

concert rooms are not far from London Zoo, and there was an initial concern that some of the animals might be affected by the infrasound, thus recreating the 'unusual uproar' initiated by Galton's sound studies over a century before. A few 'back of the envelope' calculations revealed that our four-legged friends at the zoo had nothing to worry about. However, the same calculations also showed that if we were not careful, the humans in the hall did. High levels of infrasound can inflict unpleasant effects on people's bodies. Clearly, we wanted to expose the audience only to levels that were safe. The potential problem was that as the infrasound bounced around the hall, there could possibly be areas of the auditorium where the waves combined together to create an unusually loud, and potentially dangerous, effect. To prevent this happening, it was important to turn on the pipe before the concert, and have Richard and Dan carefully sweep through the auditorium checking the infrasound levels.

The team assembled on the morning of the concerts, the pipe was installed at the back of the hall, turned on to maximum power, and the sweep began. Thankfully, the results revealed that no parts of the hall were exposed to dangerously high levels of infrasound. Relieved, we continued with our preparations.

My role was to compère the event. To welcome people, explain the purpose of the experiment, and ensure that the questionnaires were properly completed. Ciarán had decided which pieces of music would include the infrasound, and so sat with Richard and Dan as they controlled the pipe. Sarah was team leader, and also presented a talk after the concert explaining the science underpinning the event. GéNIA played each of the pieces during the concert.

Carrying out these types of live events is always nerve-racking. There is usually only one chance to get everything right, and, if anything does go wrong, there is a high potential for considerable public humiliation. Pre-publicity had ensured that the event was a sell-out, and GéNIA and I waited nervously backstage as 200 members of the public filed into the hall for the first concert. The lights in the auditorium slowly dimmed, and I walked onto the stage and welcomed people to the unique event. GéNIA played each of the pieces perfectly, the pipe was turned on and off on cue, and the audience had a thoroughly enjoyable time. Everyone completed their questionnaires at the end of the four experimental pieces of music, and handed them to us as they left the hall. I needn't have been nervous. The whole concert ran like clockwork. About an hour later we repeated the entire process for the 200 people attending the second concert, and then we retired to the nearest bar.

In the following week, my research assistant entered the questionnaire data into a computer, and analysed the results. Had all of Sarah's careful planning and preparation paid off? Had the infrasound really produced any spooky effects in our concert-going guinea pigs? If so, this would be the first experimental evidence to suggest that Vic was right to think that some alleged ghostly experiences may be due to low-frequency sound waves. The good news was that no member of the audience had experienced the dreaded 'brown note' phenomenon. The very good news was that, as predicted, they had reported significantly more strange experiences during the pieces when they incorporated infrasound. The effect was far from trivial, with people reporting, on average, about 22 per cent more unusual experiences with infrasound present. People's description of their unusual experiences made for fascinating reading. When the infrasound was flooding into the concert room during one piece, one audience member reported a 'shivering on my wrist, odd feeling in stomach', whilst another said that he had an 'increased heart rate, ears fluttering, anxious'. At another point in the concert, one man said he 'felt like being in a jet before it takes off', whilst a woman reported a 'pre-orgasmic tension in body and arms, but not in legs'.

These findings were reported by the world's media. As a result, the team was contacted by various theme parks asking whether they could use infrasound to make their scariest rides even more terrifying. This was not, however, the most curious spin-off from the project. We had shown that some 'ghostly' experiences may be due to infrasound. Some academics, however, have taken the idea one step further, suggesting that the same low-frequency waves might also play a key role in creating allegedly sacred experiences. Aeron Watson and David Keating from Reading University have constructed a computer model of a Scottish Neolithic passage grave.[42] Using this model, the researchers have argued that the site has an infrasonic resonant frequency, such that a person beating a 30-centimetre drum could produce powerful low-frequency sounds.[43] Others have suggested that large organ pipes found in certain churches and cathedrals are capable of producing similar effects.

As part of the preparation for the concert, the team visited several churches and cathedrals that contained especially large organ pipes, and discovered that some were indeed creating significant levels of infrasound. This suggests that people who experience a sense of spirituality in church may be reacting to the extreme bass sound produced by the pipes. Further support for the idea came from one pipe manufacturer who informally told the team that, given that the sounds from these pipes are inaudible, they can be viewed as either a very expensive way of creating a small draught, or a cost-effective way of helping the congregation find God.

4

Making your mind up: The strange science of decision-making

Why incompetent politicians get elected and the innocent
are found guilty, how to create the perfect chat-up line and
personal ad, whether subliminal perception affects buying
behaviour, and why John and Susan Fish *really* wrote
A Student's Guide to the Seashore.

So here's the deal. Imagine that you decide to buy a nice, new, calculator. You go to the calculator shop, and the assistant shows you a range of devices. After careful consideration, you choose a model that costs £20. At this point the assistant looks slightly anxious, and explains that the following day the shop is going to have a sale. If you come back then, the calculator will cost only £5. Do you buy the calculator then and there, or return the following day?

Now let's imagine a slightly different scenario. This time you decide to buy a new computer. You go in, and the assistant shows you a range of machines. After careful consideration, you choose a computer costing £999. Once again, the assistant looks anxious, and explains that the following day there will be a sale. If you come back then, the computer will be reduced to £984. Do you buy the computer, or return the following day?

Researchers examining the psychology of decision-making

have presented these two scenarios to lots of people. In both instances, people have the opportunity to save identical amounts of money, and so it would be rational to treat them in the same way. People should either buy the calculator and computer straight away or, if they want to save £15, return to the shops the following day. Most people treat the two scenarios very differently, however. About 70 per cent of people say that they would put off buying the calculator until the following day, but purchase the computer right there and then.

Even without a calculator, it is clear that the figures don't add up. Why do so many people act in such an irrational way? It seems that they don't view their potential saving in absolute terms, but rather as a percentage of the amount of money they are spending. In absolute terms, each time they stand to save £15. However, this represents 75 per cent of the price of the calculator, but just 1.5 per cent of the price of the computer. Seen in relative terms, the former seems to be a much better deal than the latter, and so well worth waiting for.

This is just one example of the large amount of research investigating how people make up their minds. The work has examined how people make many different types of decisions, including who they should marry, which political party they should support, the type of career they wish to pursue, the sort of house they should live in, the size of car they should have, and whether they should give it all up and move to the country.

We are going to explore the unusual work that has been carried out on decision-making. Whether subliminal messages can increase sales of Coke, popcorn, and bacon; whether something as simple as the height of political candidates can cause voters to switch allegiance from one party to another. Does your surname influence where you live and the career you pursue? Do Hollywood movies influence the verdicts returned in courtrooms across the globe? And why are certain chat-up lines and personal ads more effective than others?

We start by delving deep into the strange world of subliminal perception.

Drinking Coke, eating popcorn, and buying bacon

In September 1957, market researcher James Vicary announced the results of a remarkable experiment proving that sublim-inal stimuli exerted a powerful influence over people's buying behaviour.[1] According to Vicary, cinemagoers in New Jersey had been secretly exposed to the subliminal messages 'Drink Coke' and 'Eat popcorn' whilst watching their favourite films. These messages had been flashed on the screen using a high-speed projector designed by Vicary, with each exposure lasting just one three-thousandth of a second. Although the audience had been unaware of the messages, sales of Coke and popcorn had increased by 18 per cent and 58 per cent respectively. Vicary's announcement generated a considerable furore among the public and politicians. Could people's thoughts and behaviour really be manipulated by subliminal messages? Could people be persuaded to buy products they didn't want, and vote for politicians they didn't support? Could these messages even be broadcast on national television and influence an entire nation?

Word about the possible power of subliminal stimuli

spread like wildfire, with a survey conducted just nine months after Vicary's press conference revealing that over 40 per cent of respondents had heard about the story. The resulting hullabaloo caught the attention of Melvin DeFleur, an expert in communication studies from Indiana University. DeFleur had gained his doctorate for CIA-funded research examining how information about food and shelter could be effectively communicated to the public in the event of a nuclear war.[2] DeFleur had been especially interested in the effectiveness of two rather low-tech methods of dissemination – word of mouth and the bombarding of towns with thousands of leaflets. To avoid the risk of causing widespread panic, DeFleur and his colleagues often disguised the real reason they were carrying out their work. In one part of the project, researchers visited a fifth of homes in an isolated town in Washington State posing as marketers for the Gold Shield Coffee Company. They told people that the company had developed a new slogan ('*Gold Shield Coffee – Good as Gold*'), and that in three days' time they would interview all of the inhabitants in the town, and give a pound of coffee to everyone who could remember the slogan. In addition to this face-to-face attempt to create a caffeine-related buzz, the American air force were ordered to bombard the town with 30,000 leaflets describing the scheme. When the investigators arrived three days later, they discovered that 84 per cent of inhabitants were able to tell them accurately that Gold Shield Coffee was as good as gold. In their report, the researchers note that this figure may represent an unrealistically high level of dissemination because the price of coffee had risen dramatically just before the start of the study, and the public may have been highly motivated to remember the slogan.

DeFleur was curious about the claims being made by James Vicary concerning subliminal perception, and teamed up with colleague Robert Petranoff to investigate.[3] The two decided to conduct a realistic test by presenting hidden messages on national television. They knew that they had to be quick. The National Association of Broadcasters had already recommended that subliminal stimuli should not be used on the media, and it seemed likely that a full ban was on its way. DeFleur and Petranoff carried out two experiments on the television station WTTV Channel 4, in Indianapolis.

The first part of the study was designed to determine whether hidden messages could affect the public's viewing habits. As part of its normal nightly programming, WTTV Channel 4 broadcast a two-hour feature film, followed by a news programme hosted by a well-known presenter named Frank Edwards. The experimenters obtained permission to superimpose the subliminal message 'Watch Frank Edwards' throughout the entire two-hour film, in the hope that it would persuade more people to tune into the Edwards show.

A second aspect of the experiment examined the possibility that subliminal stimuli might alter people's buying behaviour. John Fig, Inc., a wholesale bacon distributor in Indiana, allowed the experimenters to flash the subliminal message 'Buy bacon' during its television commercials, and then to track the resulting effect on sales across the region.

Throughout July 1958, people watching WTTV Channel 4 were bombarded with hidden messages that told them to watch Frank Edwards and buy bacon. Before the experiment, an average 4.6 per cent of the public had been tuning in to Frank Edwards. After being exposed to two hours of continual subliminal messages, that figure fell to just 3 per cent. The effect of the subliminal messages on buying behaviour were just as unimpressive. Before the experiment, John Fig, Inc. sold an average of 6,143 units of bacon per week to the good folk of Indiana. By the end of the study, the figure had shown a very modest increase to 6,204 units per week. In short, the subliminal stimulation had had almost no effect on bacon sales, and, if anything, had persuaded a considerable number of people to avoid Frank Edwards. The effects of the subliminal onslaught had been less than remarkable.

DeFleur and Petranoff concluded that the public could sleep easily at night, safe in the knowledge that they were not having their thoughts and behaviour secretly manipulated by subliminal stimuli.

They were not the only ones to investigate the issue. A few months before, the Canadian Broadcasting Company briefly presented the phrase 'Phone now' more than 350 times during a popular Sunday night programme called *Close-up*, and asked viewers to write in if they noticed any strange change in their behaviour. CBC saw no significant increase in telephone usage during, or after, the programme. The station did, however, receive hundreds of letters from viewers describing how they had experienced an unaccountable urge to drink beer, visit the bathroom, or take their dog for a walk. But despite the impressive lack of evidence suggesting that televised subliminal stimuli had any effect on viewers, the National Association of Broadcasters responded to public and political pressure in June 1958 by banning the use of these messages on American networks.

So why the discrepancy between the increase in sales of popcorn and Coke claimed by James Vicary, and the lack of bacon-buying reported by DeFleur and Petranoff? The mystery was finally resolved in 1962, when Vicary was interviewed in the magazine *Advertising Age*. He explained how his story about subliminal stimuli and buying behaviour had been leaked to the media far too early. In fact, he had only collected the minimum amount of data needed to file a patent, and admitted that his investigations were far too small to be meaningful. The entire public and political debate had been based on fiction, not fact. Towards the end of his interview, Vicary added, 'All I accomplished, I guess, was to put a new word into common usage . . . I try not to think about it anymore.' Vicary did far more than simply encourage people to use the word 'subliminal'. His fictitious study has become the stuff of urban legend, and is still referred to by those who believe that buying behaviour can be influenced by subliminal messages.

The lack of evidence to support a link between televised subliminal messages and behaviour has not stopped present-day politicians worrying about the possible effect of subtle signals on voters. During the 2000 US presidential elections, the Republicans produced a television advertisement criticizing the Democrats' policy towards prescription drugs for the elderly. As part of the advertisement, various words slowly moved from the foreground to the background. As the word 'bureaucrats' came into view, one frame of the advertisement contained just the last four letters of the word, spelling 'rats'. The Democrats perceived this as an attempt to sway the electorate via subliminal perception, and asked the Federal Communications Commission to investigate the matter. The Republicans dismissed the appearance of the 'rats' word as coincidence, and argued that the advertisement was about health care and not rodents.

James Vicary is not the only person to claim that subliminal stimuli can exert a powerful effect over people's behaviour. Others have written best-selling books claiming that advertisers regularly implant sexually arousing images in photographs to help boost sales. Alleged examples include women with bare breasts in ice cubes, a man with an erection on cigarette packs, and the word 'sex' embedded several times on each side of one of the world's most popular biscuits. In addition, several companies have marketed subliminal audiotapes containing hidden messages that claim to produce all sorts of desirable effects, including increased self-esteem, sexual prowess, and intelligence. This is not small business. In 1990, it was estimated that sales of subliminal audiotapes exceeded $50 million annually in America alone.[4] Most of these claims have not been subjected to any form of scientific testing, and the few studies that have been carried out into the topic have obtained null results.[5] In one study, overweight people listened to subliminal audiotapes designed to help them drop a size. They lost no more weight than a control group not listening to any tapes.[6] In another experiment, police officers spent over twenty weeks listening to tapes designed to improve their marksmanship.[7] The results revealed that the group ended up with same shooting abilities as their non-subliminally stimulated colleagues.

So, does this mean that our thinking and behaviour isn't influenced by small, subtle, signals? In fact, there is a large amount of research to suggest that many aspects of our everyday behaviour *are* affected by factors outside of our awareness. These factors are not to be found briefly flashed up on cinema and television screens. Instead, they are right in front of our noses, and can exert a considerable influence on the way we think and behave – like something as simple as your name.

Mr Bun the baker

In 1971, psychologists Barbara Buchanan and James Bruning had a group of people rate how much they liked over a thousand first names.[8] Strong stereotypes emerged, with the vast majority of people giving the thumbs up to the likes of *Michael*, *James*, and *Wendy*, but showing a strong dislike for *Alfreda*, *Percival*, and *Isidore*. It would be nice to think that these emotional reactions don't have a significant effect on people's lives. Nice, but wrong.

In the late 1960s, American researchers Arthur Hartman, Robert Nicolay and Jesse Hurley examined whether people with unusual names were more psychologically disturbed than their normally named peers.[9] They examined more than 10,000 psychiatric court records, and identified eighty-eight people with highly unusual first names, such as Oder, Lethal, and Vere. They then looked through the same set of records and put together a control group of eighty-eight normally named people who were matched on gender, age, and place of birth. Those with unusual names were significantly more likely than the control group to be diagnosed as psychotic. As the researchers note in their paper, 'A child's name . . . is generally a settled affair when his first breath is drawn, and his future personality must then grow within its shadow.' This study is not alone in documenting the downside of having a name that stands out from the crowd. Research has shown that teachers award higher essay grades to children with likeable names,[10] that college students with undesirable names experience high levels of social isolation, and that people whose surnames happen to have negative connotations (such as 'Short', 'Little' or 'Bent') are especially likely to suffer feelings of inferiority.[11] American psychiatrist William Murphy has examined several case histories illustrating this final point. In one instance, a patient admitted to wearing an athletic support to bed when he was a boy to prevent his penis becoming erect. The support failed to have the desired effect, and instead caused his penis to bend downwards. Unfortunately, the patient's last name was Bent, and this, coupled with the fact that his nickname was 'Dinkey', constantly reminded him of the sexual problems that he had experienced as a boy. This, in turn, made him feel anxious about sex, resulting in psychosexual impotence and reinforcing his feelings of inadequacy.

Nicholas Christenfeld, David Phillips and Laura Glynn from the University of California, San Diego, uncovered evidence in 1999 suggesting that even a person's initials may become an issue of life or death.[12] The team used an electronic dictionary to generate every three-letter word in the English language. They then worked their way through the list, identifying words that were especially positive (such as ACE, HUG and JOY) and those that had very negative connotations (PIG, BUM and DIE). Using a computerized database of Californian death certificates, they examined the age at which people with 'positive' and 'negative' sets of initials passed away. Controlling for factors such as race, year of death, and socio-economic status, the researchers discovered that men with positive initials lived around four and a half years longer than average, whereas those with negative initials died about three years early. Women with positive initials lived an extra three years, although there was no detrimental effect for those with negative initials. When discussing the possible mechanisms behind the effect, the authors noted that people with negative initials 'may not think well of themselves, and may have to endure teasing and other negative reactions from those around them'. This idea was supported by the fact that those with negative initials were especially likely to die from causes with psychological underpinnings, such as suicides and accidents.

But it is not all doom and gloom for the unusually named and negatively initialled. For a start, another team of researchers have questioned the findings from the Christenfeld study. In a paper entitled 'Monogrammic Determinism?', Stilian Morrison and Gary Smith from Pomona College in California criticize the statistical methods used in the original experiment, and failed to replicate the findings using what they consider to be more sophisticated analyses.[13]

Furthermore, psychologist Richard Zweigenhaft, from Guilford College in North Carolina, has argued that there are several potential benefits associated with having an unusual name.[14] He notes that one of the most frequently voiced complaints by those with common names is that there are too many other people with the same name. The same point was well made by Samuel Goldwyn, who, upon hearing that a friend had named his son John, quipped, 'Why did you name him John? Every Tom, Dick, and Harry is named John.' Zweigenhaft also notes that unusual names are more memorable, and cites several instances in which the fame enjoyed by well-known sportspeople may have been due, at least in part, to their unusual name. As one *New York Post* sportswriter noted when discussing the Oakland Athletics' pitcher Vida Blue, 'America knew it instantly. Vida Blue! Vida Blue tripped off the tongue like Babe Ruth and Ty Cobb and Lefty Grove.'

Taking a more empirical look at the potential positive effects of unusual naming, Zweigenhaft randomly selected 2,000 people from the *Social Register* (described as the 'best guide to the membership of the national upper classes'), and identified names that were mentioned only once. This process generated a list of 218 people. Zweigenhaft then generated a control group by randomly selecting 218 people who did not have unusual names in the original sample of 2,000. Next, he consulted *Who's Who* (described as a book listing 'the best-known men and women in all lines of useful and reputable achievement'), to discover whether people with usual or unusual names tended to attain eminence. Of the total of 436 possibilities (2 × 218), thirty appeared in *Who's Who*. Twenty-three of these were from the 'unusual names' group listed in the *Social Register*, versus just seven of those with more usual names. In short, evidence that under certain circumstances, an unusual name can be good for your career.

Work examining the effect that people's names have on their life is not just concerned with whether a name is unusual or usual. The remarkable research of Professor Brett Pelham and his colleagues at the State University of New York at Buffalo suggests that our names may influence the towns in which we choose to live, the career paths we follow, the person we marry, and even the political parties we support.[15]

By looking at a huge number of American census records, Pelham has uncovered an over-representation of people called Florence living in Florida, George in Georgia, Kenneth in Kentucky, and Virgil in Virginia. In another study, the team examined the social security death records of 66 million American people who had died in cities starting with the word *Saint* (for example, St Anne, St Louis, etc.). Once again, they found proportionately more people called Helen in St Helen, more Charleses in St Charles, more Thomases in St Thomas, and so on. Further analyses suggested that these effects are not owing to people naming their offspring after their place of birth, but rather to people drifting towards cities and towns containing their own name.

Could the same effects even influence people's choice of marriage partner? Are people more likely to marry someone whose surname starts with the same letter as their own? To find out, Pelham and his colleagues looked at over 15,000 marriage records between 1823 and 1965.[16] An intriguing pattern emerged, with significantly more couples sharing the initial letter of their family name than predicted by chance. Worried that the effect might be due to ethnic matching (that is, members of certain ethnic groups being likely to marry one another and have surnames starting with certain letters), the team repeated the study, but this time focusing their attention on the five most common American surnames – Smith, Johnson, Williams, Jones, and Brown. Once again the effect emerged, with, for instance, people named Smith being more likely to marry another Smith than someone called Jones or Williams, and people called Jones more likely to say 'I do' to another Jones than a Brown or even a Johnson.

Pelham's work is not restricted to examining the relationship between people's names, where they choose to live and die, and the people that they marry. He has also examined how their surnames may influence their choice of occupation. Searching the online records of the American Dental Association and American Bar Association, the researchers found that there were more dentists whose first names start with the three letters 'Den', than the three letters 'Law'. Likewise, there was a greater preponderance of lawyers whose first names start with the three letters 'Law', than 'Den'. Then there is the data from hardware and roofing companies. Using the Yahoo Internet Yellow Pages, the team searched for all of the hardware stores and roofing companies in the twenty largest American cities, and examined whether the owners' first name or surname began with the letter 'H' or 'R'. The results revealed that the names of owners of hardware companies tended to start with the letter 'H' (such as Harris Hardware), whilst the names of those in charge of roofing companies tended to start with 'R' (such as Rashid's Roofing). According to Pelham, the same effect even extends into politics. During the 2000 presidential campaign, people whose surnames began with the letter 'B' were especially likely to make contributions to the Bush campaign, whereas those whose surnames began with the letter 'G' were more likely to contribute to the Gore campaign. Writing about his results in a paper entitled 'Why Susie Sells Seashells by the Seashore: Implicit Egotism and Major Life Decisions', Pelham concludes that perhaps we should not be surprised by these effects, noting that they '. . . merely consist of being attracted to that which reminds us of the one person most of us love most dearly'.

In addition to being interesting in its own right, Pelham's work may at last provide an explanation for an effect that has fascinated psychologists for decades: Why does the meaning of people's surnames so often match their chosen occupation?

In 1975, Lawrence Casler from the State University of New York at Geneseo compiled a list of over 200 academics working in fields associated with their last names.[17] Casler's list includes an underwater archaeologist called Bass, a relationship counsellor called Breedlove, a taxation expert named Due, a medic studying diseases of the vulva called Hyman, and an educational psychologist examining parental pressure called Mumpower. In the later 1990s, *New Scientist* magazine asked readers to send in similar examples from their own lives. The resulting list included music teachers Miss Beat and Miss Sharp, members of the British Meteorological office called Flood, Frost, Thundercliffe and Weatherall, a sex counsellor named Lust, Peter Atchoo the pneumonia specialist, a firm of lawyers named Lawless & Lynch, private detectives Wyre & Tapping, and the head of a psychiatric hospital, Dr McNutt. My own favourites are the authors of the book *A Student's Guide to the Seashore*, John and Susan Fish.

Pelham's work suggests that examples like this may not be entirely the result of chance, but rather of some people being unconsciously drawn to occupations related to their name. As a professor of psychology called Wiseman, I am in no position to be sceptical about the theory.

Hidden persuaders

Our names are assigned to us the moment we are born, and, for most people, remain throughout our lives. However, some of the other factors that influence our thoughts and behaviour are far more subtle. Sometimes, it can just be a matter of a single sentence, short piece of music, or newspaper headline.

It really doesn't take much to change the way in which a person thinks, feels, and behaves. The concept is beautifully illustrated in two studies recently published in one of the world's most prestigious academic publications, the *Journal of Personality and Social Psychology*.

In the first of these, conducted by John Bargh and his colleagues at New York University, participants were asked to rearrange a series of scrambled words to form a coherent sentence.[18] Half of the participants were shown mixed-up sentences that contained words relating to the elderly, such as 'man's was skin the wrinkled'. The other half of the participants were shown the same mixed-up sentences, but the one word relating to the elderly was replaced with a word not associated with old age, such as 'man's was skin the smooth'. Once a participant had carefully worked their way through the sentences, and been thanked for taking part, the experimenter gave directions to the nearest set of elevators. The participant thought that the experiment was over. In reality, the important part was just about to start. A second experimenter was sitting in the hallway armed with a stopwatch. When the participant emerged from the laboratory, this second experimenter secretly recorded the time taken for them to walk down the hallway to the elevators. Those who had just spent time unscrambling the sentences that contained words relating to old age took significantly longer than those who had spent time with the non-elderly sentences. Just spending a few minutes thinking about words such as 'wrinkled', 'grey', 'bingo', and 'Florida', had completely changed the way people behaved. Without realizing it, those few words had 'added' years to their lives and they were walking like elderly people.

A similar study, conducted by Ap Dijksterhuis and Ad van Knippenberg from the University of Nijmegen in Holland, asked participants to spend five minutes jotting down a few sentences about the behaviour, lifestyle, and appearance of a typical football hooligan, whilst others did exactly the same for a typical professor.[19] Everyone was then asked about forty Trivial Pursuit questions, such as 'What is the capital of Bangladesh?', 'Which country hosted the 1990 Soccer World Cup?', and so on. Those who had spent just five minutes thinking about a typical football hooligan managed to answer 46 per cent of the questions correctly, whereas those who had generated sentences related to a typical professor were right 60 per cent of the time. Without people being aware of it, their ability to answer questions correctly was dramatically altered by them simply thinking about a stereotypical football hooligan or professor.

This is all well and good within the relatively artificial confines of a laboratory, but do the same effects influence people's behaviour in the real world?

Americans leave about twenty-six billion dollars in restaurant tips every year. You would think the size of tip depends on the quality of food, drink, or service provided, but secret studies conducted in bars and restaurants across the globe have revealed the hidden factors that really determine our tipping behaviour. Mood plays a large part in the process. Happy eaters are bigger tippers. In one study, French bar staff were asked to give their customers a small card with the bill.[20] Half of the cards contained an advertisement for a local nightclub, whilst the other half contained the following joke:

An Eskimo had been waiting for his girlfriend in front of a movie theatre for a long time, and it was getting colder and colder. After a while, shivering with cold and rather infuriated, he opened his coat and drew out a thermometer. He then said loudly, 'If she's not here by 15 degrees, I'm going!'

Those receiving the joke showed a higher level of laughing and, more importantly, tipping. Researchers have replicated the relationship between happiness and tipping time and again. Waiters get bigger tips when they draw happy faces, or write 'Thank You' at the bottom of a bill, or give a big smile to customers.[21] People tip more when the sun is shining, and even when waiters tell them that the sun is shining.[22] Other studies have shown that tipping is dramatically increased when waiters introduce themselves using their first name, or refer to customers by their name.[23]

Then there is the power of touch. Describing their work in a paper entitled 'The Midas Touch: The Effects of Interpersonal Touch on Restaurant Tipping', April Crusco explains how she trained two waitresses to touch the diners' palm or shoulder for exactly one and a half seconds as they gave them the bill.[24] Both kinds of touching produced more tipping than the hands-off approach adopted in the control condition, with palm-touching doing slightly better than a tap on the shoulder.

Leaving relatively small amounts of money to waiters and bar staff is one thing, but can these subtle effects persuade people to part with much larger sums of cash?

In the 1990s, researchers Charles Areni and David Kim from Texas Tech University investigated exactly this issue by systematically varying the music being played in a downtown wine shop.[25] Half of the customers were subjected to classical tunes, including Mozart, Mendelssohn, and Chopin, whilst the other half heard pop songs, including Fleetwood Mac, Robert Plant, and Rush. By disguising themselves as shop assistants making an inventory of stock, the experimenters were able to observe customers' behaviour, including the number of bottles they picked up from the shelves, whether they read the labels, and, most important of all, the amount of wine they bought. The results were impressive. The music did not affect how long people stayed in the cellar, how many bottles they examined, or even the number of items bought. Instead, it had a dramatic effect on just one aspect of their behaviour – the cost of the wine they bought. When the classical music was playing, people bought bottles of wine that were, on average, over three times more expensive than when the pop music was playing. The researchers believe that hearing the classical music unconsciously made them feel more sophisticated, and that this, in turn, caused them to buy significantly more expensive wine.

There is even some evidence to suggest that the same sort of subtle stimuli influence matters of life and death.

An analysis of over 1,400 country songs by sociologist Professor Rogers (Jimmie, not Kenny) revealed that the lyrics often refer to negative life experiences, including unrequited love, alcohol abuse, financial problems, hopelessness, fatalism, bitterness, and poverty.[26] In the mid 1990s, Steven Stack from Wayne State University and Jim Gundlach at Auburn University wondered whether continual exposure to downbeat topics might make people more likely to commit suicide.[27] To find out, the researchers looked at the suicide rate, and the amount of country music played on national radio, in forty-nine areas across America. After controlling for several other factors, such as poverty, divorce, and gun ownership, the researchers did find that the more country music played on radio, the higher the suicide rate.

The results may sound far-fetched, and have been challenged by several other researchers.[28] The basic premise, however, is supported by a wealth of other work suggesting that the mass media plays an important role in determining whether people decide to end their lives, of which the study of the 'Werther Effect' is an excellent example.

In 1774, Johann von Goethe published a novel entitled *The Sorrows of Young Werther*. In the book, a young man named Werther falls in love with a woman who is already promised to another. Rather than face a life without her, Werther decides to end his life by shooting himself. The book was a remarkable success. In fact, in many ways it was a little too successful, inspiring a series of copycat suicides that eventually resulted in it being banned in several European countries. In 1974, sociologist David Phillips, from the University of California at San Diego, decided to examine whether media reports of suicides might create a modern-day Werther effect.[29] In an initial study examining the suicide statistics across America between 1947 and 1968, he discovered that a front-page suicide story was associated, on average, with an excess of almost sixty suicides. Moreover, the type of suicides reflected the method of death outlined in the media, and the level of publicity received by the suicide was directly related to the number of subsequent deaths. On average, the number of suicides increased by roughly 30 per cent within two weeks of media reports, and the effect was especially pronounced after a celebrity death. Phillips calculated, for example, that the death of Marilyn Monroe in August 1962 increased the national suicide rate by about 12 per cent. Since Phillips' groundbreaking work, there have been more than forty scientific papers on the topic, prompting several countries to produce media guidelines urging reporters not to sensationalize suicides, or describe the methods that people used to kill themselves.[30]

Another part of Phillips' work has investigated the relationship between televised boxing matches and murder rates. He carefully analysed daily murder rates across America, and showed that they tended to increase in the week following the television broadcast of a high-profile heavyweight boxing match. Not only was there a direct relationship between the amount of publicity the fight received and the number of murders, but also between the racial backgrounds of the boxers and the murder victims. If a white boxer lost the fight, Phillips found an increase in the number of white, but not black, people murdered. Likewise, if a black boxer lost, there was an increase in the number of black, but not white, people killed.

All of this adds up to one simple fact. The ways in which we think and feel are frequently influenced by factors outside our awareness. Our names influence our self-esteem and choice of career. Just reading a sentence can influence how old we feel and our recall of general knowledge. A simple smile or subtle touch can influence how much we tip in restaurants and bars. The music played in shops creeps into our unconscious minds, and influences the amount of money we spend. But do the same sort of strange persuaders also influence the way in which we see others? Could they even dictate the politicians that we vote for, and the way in which we decide upon the guilt or innocence of our fellow citizens?

Inching forward in the polls

Thousands of years ago there were evolutionary advantages to hanging around with taller people because their physical size afforded all sorts of benefits when it came to gathering food and defeating foes. Although height no longer offers any physical advantage, our primate brains hold on to their evolutionary past, and so still associate tall people with success: a faulty but persuasive perception that plays out in various ways.

Psychologists Leslie Martel and Henry Biller asked university students to rate men of varying heights on many different psychological and physical attributes.[31] Reporting the results in their book, *Stature and Stigma*, they describe how both men and women rated men under five foot five as less positive, secure, masculine, successful, and capable. Even our language reflects the value of height. Those held in high esteem are 'big men', that we 'look up' to. Run out of money, and you are 'short' of cash.

Even in the world of romance and mating, size matters. Professor Dunbar, an evolutionary psychologist at Liverpool University, and his colleagues analysed data from over 4,000 healthy Polish men who had had compulsory medical examinations between 1983 and 1989.[32] They found that childless men were about three centimetres shorter than men who had at least one child. The only exceptions to the pattern were men born in the 1930s. Dunbar believes that this was because they emerged into the marriage market just after the Second World War, when single men were relatively scarce and so women had little choice.

This association between mating success and height appears to be universal. In the 1960s, anthropologist Thomas Gregor from America's Vanderbilt University lived among a tropical forest people of central Brazil known as the Mehinaku.[33] Even here, height matters. Among the Mehinaku, tall men are seen as attractive and are respectfully referred to as 'wekepei'. Those short in stature are referred to by the derisive term 'peritsi', which rhymes with *itsi*, the word for penis. Wekepei were far more likely than peritsi to be asso-ciated with wealth, power, participation in rituals, and reproductive opportunities. Gregor discovered that the taller the man, the more female mates he had access to, with the three tallest men having had as many affairs as the seven shortest men.

Does height also matter when it comes to careers? It seems so. In the 1940s, psychologists found that tall salesmen were more successful than their shorter colleagues, and a 1980 survey found that more than half the CEOs of America's Fortune 500 companies were at least six feet tall. More recent research from the *Journal of Applied Psychology* shows that when it comes to height in the workplace, every inch counts.[34] Business Professor Timothy Judge, of the University of Florida in Gainesville, and his colleague Daniel Cable analysed the data from four large studies that had followed people through their lives, carefully monitoring their personality, height, intelligence, and income. Focusing on the relationship between height and earnings, Judge discovered that each inch above average corresponds to an additional

$789 in pay each year. Someone who is six feet tall therefore earns an extra $4,734 more each year than their equally able five-foot-five colleague. Compounded over a thirty-year career, a tall person enjoys an earning advantage of hundreds of thousands of dollars over shorter colleagues.

Politics has also been scrutinized. Of the forty-three American presidents, only five have been below average height, and it has been over one hundred years since voters elected someone who was shorter than average (the five-foot-seven-inch-tall President William McKinley, who took office in 1896, and was referred to by the press as a 'little boy'). Most presidents have been several inches above the norm. Ronald Reagan was six foot one, and George Bush Senior and Bill Clinton both stand tall at six foot two. There is also evidence to suggest that some candidates recognize the importance of height with voters, and take steps to make the most of any advantage. In the 1988 presidential debate, George Bush Senior greeted Michael Dukakis with an exaggerated long handshake – a move apparently orchestrated by Bush's campaign manager to make the most of the fact that Bush was taller.

The psychological relationship between status and height works in both directions. Not only do we think that tall people are more competent, but also that competent people are tall. This explains why people are often surprised to discover that some Hollywood stars are below average height. Dustin Hoffman, for example, is just five foot five, and Madonna is five foot four inches tall. The website www.celebheights.com (byline: 'In the land of Hollywood Pygmies, the elevator-shoed Dwarf is King') is dedicated to discovering the true heights of celebrities, often sending people of a known height to have their photograph taken next to celebrities in order that their heights can be accurately determined. The author Ralph Keyes speculated about the fact that so many actors are short in his book *The Height of Your Life*. Keyes thought that some smaller people have a need to show that they are strong and overcome their height disadvantage by developing strong personalities.

This relationship leads to an interesting phenomenon – that the perceived height of a person can change with their apparent status. The first scientifically controlled experiment into this curious phenomenon was conducted by psychologist Paul Wilson from the University of Queensland.[35] Wilson introduced a fellow academic to different groups of students and asked them to assess his height. Unbeknown to the students, Wilson changed the way in which he introduced the academic each time. On one occasion he told the class that the man was a fellow student, the next time he said that he was a lecturer, then the man was introduced as a senior lecturer, and finally as a full professor. The students' perception of the man's height varied with his perceived status. When he was just a fellow student he was seen as being about five feet eight inches tall. However, simply saying that he was a lecturer added about an inch to his height. Promoting him to senior lecturer meant that he gained another inch in the eyes of the students, whilst his rather rapid promotion to professor brought him up to about six foot.

In 1960, Harold Kassarjian from the University of California asked 3,000 voters whether they would be supporting Kennedy or Nixon in the forthcoming election, and which they believed to be the taller of the two candidates.[36] In reality, Kennedy was an inch taller than Nixon. However, this was not how his voters saw it. Forty-two per cent of Nixon supporters said that Nixon was the taller candidate, compared to just twenty-three per cent of Kennedy supporters. Other research, conducted in the early 1990s by Philip Higham and William Carment at McMaster University in Canada, took things a stage further.[37] Higham and Carment asked voters to estimate the heights of the leaders of the three main political parties in Canada (Brian Mulroney, John Turner, and Ed Broadbent) before, and after, a general election. Mulroney won the election, resulting in a half-inch gain in the height polls. After losing the election, Turner and Broadbent were seen to have shrunk by about a half an inch, and one and a half inches, respectively.

I wondered whether it might be possible to use this effect to measure the perceived status of politicians before an election. In 2001, I worked with Roger Highfield, science editor at the *Daily Telegraph*, to carry out an unusual political opinion poll.[38] We asked a representative sample of 1,000 respondents to estimate the height of the leaders of the UK's two main political parties. According to their party headquarters, the Labour and Conservative leaders at that time, Tony Blair and William Hague, are both six feet tall. But this is not how the electorate saw things.

In line with Harold Kassarjian's findings from the 1960s, we found differences when people estimated the heights of the leader they supported and the leader they opposed. Significantly more Labour than Conservative voters thought that Blair was five foot nine or over. Likewise, more Conservatives than Labour supporters thought that Hague was five foot nine or above. In short, voters saw their own candidates as taller than the opposition. However, what did our stature poll predict about the results of the forthcoming election? Whereas only 35 per cent of voters thought that Blair was less than the average male height of five foot nine, a massive 64 per cent of voters thought this of William Hague. So voters perceived Blair as relatively tall, and Hague as a real shorty.

And the results of the 2001 election?

A massive landslide victory for Tony Blair's Labour Party.

If the face fits

Height is not the only strange factor to exert a significant influence over the way in which we view others.

We all used to be a lot hairier than we are right now. As apes, we were covered in facial and body hair, but, over the course of tens of thousands of years, we have shed our fur. There is considerable debate about why this happened. Some researchers believe that it was a result of our needing less hair to keep warm as we ventured away from the shady forests and out into the hot savannahs. Others have suggested that lack of body and facial hair was associated with a lower incidence of disease-carrying ticks and parasites. Either way, some men choose to turn back the hands of evolutionary time, and sport various types of facial hair. In doing so, they are unconsciously altering the way in which they are perceived by the people around them.

In 1973, psychologist Robert Pellegrini studied the effects of facial hair on perceived personality.[39] He managed to find eight full-bearded young men who were happy to have their facial hair removed in the name of science. Pellegrini took a photograph of each of the men before the experimental barber managed to get at them. Next, each had a photograph taken when they had a goatee and moustache, then just a moustache, and finally when they were clean-shaven. Groups of randomly selected people were asked to rate the personality of the men in the photographs. There was a positive relationship between the amount of beard, and adjectives such as masculine, mature, dominant, self-confident, and courageous. Pellegrini noted that '. . . it may well be that inside every clean-shaven man there is a beard screaming to be let out. If so, the results of the present study provide a strong rationale for indulging that demand.'

Pellegrini's work, although insightful, failed to ask about one important trait: honesty. Had he done so, his conclusions about beards might not have been so positive. Recent surveys show that over 50 per cent of the Western public believe clean-shaven men to be more honest than those with facial hair. Apparently beards conjure up images of diabolical intent, concealment, and poor hygiene. Although there is absolutely no relationship between honesty and facial hair, the stereotype is powerful enough to affect the world – perhaps explaining why everyone on the Forbes 100 List of the world's richest men is clean-shaven, and why no successful candidate for the American presidency has sported a beard or moustache since 1910.

The beard studies are just one tiny aspect of a large amount of research conducted by psychologists into the effects of facial appearance on perceived personality and abilities.

According to research recently published by Alexander Todorov and his colleagues at Princeton University, facial appearance is vitally important in politics.[40] Todorov presented students with pairs of black-and-white photographs containing headshots of the winners and runners-up for the US Senate in 2000, 2002, and 2004. From each pair of photographs, Todorov asked the students to choose which of the pair looked more competent. Even though the students only saw the pairs of photographs for just one second, choosing which of the two looked most competent predicted the actual election results about 70 per cent of the time. Not only that, but the degree of disagreement between the students also predicted the margin of victory. When the students all agreed on which of the two candidates appeared the most competent, that candidate emerged the clear winner at the polls. When there was less agreement between the students, the election results were not so clear-cut.

If facial stereotypes can influence winning or losing at the ballot box, are there other situations where looks matter? Could these same types of stereotypes even influence how people determine the guilt of defendants in a courtroom?

And so I ask the jury . . . is that the face of a mass murderer?

In chapter 2, I described how my first mass-media psychology experiment with the help of Sir Robin Day had explored the psychology of lying. Three years later, I went back to the same studio to conduct a second study. This time the experiment was bigger, and far more complicated, than before. This time, we wanted to discover whether justice really is blind.

The idea for the study had occurred to me when I had come across a Gary Larson *Far Side* cartoon. The cartoon is set in a courtroom, and the lawyer is talking to the jury. Pointing at his client, the lawyer says, 'And so I ask the jury . . . is that the face of a mass murderer?' Sitting in the dock is a man wearing a suit and tie, but instead of a normal head, he has the classic 'smiley' face consisting of just two black dots for eyes and a large semicircle grin. Like all good comedy, Larson's cartoon made me laugh and then made me think.

The decisions made by juries are serious, and so it is important that they are as rational as possible. I thought it would be interesting to put this alleged rationality to the test. During a live edition of the BBC's leading science programme, *Tomorrow's World*, the public were asked to play the role of a jury member. They were shown a film of a mock trial and had to decide whether the defendant was guilty or innocent, and then record their decision by telephoning one of two numbers.

Unbeknown to them, we would cut the country into two huge groups. We discovered that the BBC broadcasts to the nation via thirteen separate transmitters. Usually they all carry an identical signal, so that the whole country watches the same programme. For the experiment, however, we obtained special permission to send out different signals from the transmitters, allowing me to split Britain in two, and to broadcast different programmes to each half of the country.

Everyone saw exactly the same evidence about a crime in which the defendant had allegedly broken into a house and stolen a computer. However, half of the public saw a defendant whose face showed characteristics of the stereotype of a criminal – he had a broken nose and close-set eyes. The other half saw a defendant whose face matched a stereotypical innocent person – he was babyfaced and had clear blue eyes. To ensure that the experiment was as well controlled as possible, both defendants were dressed in identical suits, stood in exactly the same position in the dock, and had the same neutral expression on their face.

We carefully scripted a judge's summing-up, describing how the defendant had been accused of a burglary. The evidence presented did not allow for a clear 'guilty' or 'not guilty' decision. For example, the defendant's wife said that he was in a public house at the time of the crime, but another witness saw him leave about thirty minutes before the burglary. A footprint at the scene of the crime matched the defendant's shoes, but these were a fairly common brand of shoe owned by many people.

After transmission, we stood anxiously by the telephones, waiting to see how many calls we would receive. The experiment obviously struck a chord with the public. For the lying experiment we had received about 30,000 calls. This time, more than twice as many people telephoned. A fair and rational public would have focused solely on the evidence when deciding guilt or innocence, but the unconscious tendency to succumb to the lure of looks proved too much. About 40 per cent of people returned a guilty verdict on the man who just happened to fit the stereotype of a criminal. Only 29 per cent found the blue-eyed babyfaced man guilty. People ignored the complexity of the evidence and made up their minds on the basis of the defendants' looks.

It would be nice to think that this result is confined to the relatively artificial setting of a television studio. That, however, is not the case. Psychologist John Stewart from Mercyhurst College in Arizona spent hours sitting in courts rating the attractiveness of real defendants.[41] He discovered that good-looking men were given significantly lighter sentences than their equally guilty, but less attractive, counterparts.

In his book *Influence*, psychologist Robert Cialdini makes a fascinating link between this work and a highly unusual experiment exploring the use of plastic surgery in prisons.[42] In the late 1960s, a group of prisoners in a New York City jail were given plastic surgery to correct various facial disfigurements. Researchers discovered that these prisoners were significantly less likely to return to prison than a control group of prisoners with uncorrected facial disfigurements. The degree of rehabilitation, such as education and training, did not seem to matter. Instead, looks appeared to be everything. This result caused some social policy makers to argue that societal stereotyping was causing some people to turn to a life of crime, and that changing the way they looked had provided an effective way of preventing them reoffending. This may have been the case. However, Cialdini used the data obtained by James Stewart to argue for another interpretation of the results. It was possible that the corrective surgery had little effect on whether they reoffended, but simply meant that they were less likely to be sent to prison.

The hidden influence of Hollywood

Research shows that we link looks with likeability. Whenever we see an attractive face, we unconsciously associate it with traits such as kindness, honesty, and intelligence. Good-looking people are more likely to be offered jobs than their ugly competitors and obtain higher salaries than their equally competent colleagues.

But where do these irrational effects come from and why do they persist? Some researchers place the blame firmly at the door of Hollywood. Stephen Smith from North Georgia College and his colleagues decided to discover if this was true. In the first of two highly revealing experiments, the researchers collected twenty of the top-grossing films for each decade between 1940 and 1989.[43] They then asked a group of people to watch the films, rating all of the characters identified by name on various scales, including how attractive they were, how moral, intelligent, friendly, and whether they lived happily ever afterwards. After sitting through everything from *It's A Wonderful Life* to *Around The World In 80 Days* (the 1956 version), and from *Last Tango In Paris* to *Beetlejuice*, the raters evaluated 833 characters. The researchers discovered that physically attractive characters were depicted as more romantically active, morally good, intelligent, and far more likely than others to live happily ever after.

Although interesting, this doesn't prove that such depictions actually cause stereotypical thinking. To investigate this, the experimenters carried out a second study. They chose a few films that either did, or didn't, portray attractive people in a stereotypical way. For example, *Pride Of The Yankees* relates the true-life story of the famous baseball player, Lou Gehrig. A good-looking Gary Cooper plays Gehrig, showing his success on the field, and how in the prime of his life he begins to have serious health problems, but deals with the illness with incredible dignity. At the opposite end of the spectrum came films like *Up The Down Staircase*, in which a young and spirited teacher tries to make a difference in a troubled inner-city school. Sandy Dennis, who plays the lead in the film, was a highly acclaimed actress. However, unlike stars such as Gary Cooper, she did not look like a classic Hollywood idol, and tended to stammer her way through lines.

Groups of people were asked to watch one of the films, and rate various aspects of it. Then they were asked if they would mind helping out with a second study. They were told that a nearby university wanted people to rate the qualifications of various graduate students. Each member of the group was presented with a folder containing a résumé and a photograph of a student. In reality, all of the résumés were identical, but were accompanied by one of two pictures – one showing an attractive person and the other showing an unattractive one. Those who had just seen a film like *Pride Of The Yankees* assigned especially high ratings to the attractive candidate, and especially low ratings to the unattractive candidate. The effect disappeared when the researchers examined the corresponding data from those participants who had watched a film like *Up The Down Staircase*. Just on the showing of one film, people's perception had changed significantly. Although they weren't aware of it, the stereotypes depicted in the film had seeped into their brains, and affected the way they saw others. The experiment involved just one film. It is not difficult to imagine the effects of a lifetime of watching thousands of similarly biased television programmes, advertisements, and movies.

If you were a pizza topping, what would you be?

Knowing how your thoughts, feelings, and behaviour are influenced by subtle factors allows you to use information to your own advantage.

Right now, millions of single people right across the world are desperate to find their perfect partner (or, in many cases, any partner at all). The good news is that help is at hand. For several years, researchers have been exploring how an understanding of the psychology of attraction can help budding Casanovas impress others. Like so much of the strange science described here, the work has not been carried out in laboratories, but rather out there in the real world. During speed dating events, in personal ads, and, as with our starting point, high above the Capilano River in British Columbia.

In 1974, psychology professors Donald Dutton and Arthur Aron conducted an unusual study on two bridges above the Capilano River in British Columbia.[44] One consisted of a swaying footbridge suspended about 200 feet above the rocks, whilst the other was much lower, and more solidly built. Young men walking across each of the bridges were stopped by a female experimenter posing as a market researcher. The woman asked the men to complete a simple questionnaire, and then offered them her telephone number in case they would like to find out more about her work. As predicted by the experimenters, the offer of the telephone number was not only accepted by significantly more men on the high than the low bridge, but a greater proportion of men on the high bridge subsequently telephoned the female experimenter. Why should someone's height above the Capilano River have anything to do with them accepting the telephone number of a woman, and then calling her for a chat?

Prior to the bridge study, researchers had confirmed what poets had suspected for hundreds of years. When people see someone that they find attractive, their heart starts to beat faster as their body prepares itself for potential action. Dutton and Aron wondered whether the opposite was true – that people whose hearts were beating faster would be more likely to find someone attractive. Hence the experiment on the two bridges. The precarious nature of the high, swaying bridge meant that people using this way of crossing the river had higher heart rates than those on the lower bridge. When the men on the high bridge were approached by the female market researcher, they unconsciously attributed their increased heart rate to her, rather than to the bridge. As a result, their bodies fooled their brains into thinking that they found her attractive, and so they were more likely to want her telephone number, and subsequently give her a call. In addition to showing how the body can deceive the brain, the results have an important implication for our lives. This is why, when you want someone to fall in love with you, some scholars believe that you and your date should stay away from calming New Age music, country walks, and wind chimes. Instead, your chances of success may be increased by going to a rock concert, on a roller coaster, or to see a frightening film.

The work conducted by Dutton and Aron is just one of several unusual experiments exploring the psychology of love and attraction. Other research has tackled the rather thorny issue of chat-up lines.

If you really want to impress a potential date, what is the best opening gambit? Searching the Internet certainly won't help, with the most frequently cited lines likely to depress rather than impress ('Is it hot in here or is it just you?', 'If I could rearrange the alphabet, I'd put U and I together', and 'I've lost my phone number. Can I have yours?'). To help discover the chat-up lines most likely to attract a potential partner, researchers from Edinburgh University had people rate various types of classic openings.[45]

The results showed that straight appeals for sex ('Well, hey, there, I may not be Fred Flintstone, but I bet I can make your Bed Rock!') and compliments ('So there you are! I've been looking all over for YOU, the girl of my dreams') did not play big. In fact, they were so unsuccessful that the researchers wondered why they should have evolved at all. After much head-scratching, they concluded that these lines might be 'used by men to identify sociosexually unrestricted women' (think 'tart'). Instead, lines suggesting a potential for spontaneous wit, a pleasant personality, wealth, and an appreciation of culture, were much more effective. The study was all well and good, but, as the authors themselves admit, it is one thing to tick the 'yes, that is a good chat-up line' box on an anonymous questionnaire, and quite another to respond to it in real life.

I recently teamed up with the Edinburgh International Science Festival and academic colleagues James Houran and Caroline Watt to examine the best chat-up lines and conversational topics when searching for the love of your life. The project revolved around a large-scale experimental speed dating event.

A few months before the event, we issued a media appeal for single people who wished to participate in a study exploring the science of seduction. We had about 500 replies, and invited one hundred randomly selected participants (fifty men and fifty women) to our love laboratory.

The event took place in a large beautiful ballroom at one of Edinburgh's oldest and most palatial hotels. At the start of the evening our one hundred participants arrived, and were randomly seated at one of five long tables. At each table, men were seated on one side, and women on the other. The people at four of the tables were asked to talk about a specified topic throughout all of their speed dates. We chose four of the most frequently used topics: hobbies, film, travel, and books. Our fifth table acted as a 'control', and we allowed the people there to chat about whatever they liked. As we started to play the romantic strains of Carole King, each person was asked to chat to the person opposite them. Three minutes later, their time was up and everyone was asked to rate their potential beau. Did they find them physically attractive? What was the level of 'chemistry' between the two of them? How quickly had they made up their minds? And, perhaps most importantly, would they like to meet the person again? A few moments later, everyone was paired up with a different person, and the entire procedure was repeated. Two hours, and ten speed dates, later, and the experiment was over. It proved to be a huge success, with lots of people hanging around in the bar afterwards. Some shared their telephone numbers with one another. Others exchanged bodily fluids.

The following day we entered more than 1,500 pages of data into a giant spreadsheet. Wherever two people had indicated that they would be happy to meet up again, we sent them each other's telephone numbers. Around 60 per cent of those attending walked away with the contact details of at least one other person. Some people did really well, with about 20 per cent getting the details of four others. Women proved to be about twice as picky as men, but the top-rated man and woman of the evening had a 100 per cent success rate, with all of their dates wanting to meet them again.

The various topics of conversations had produced different success rates. When talking about movies, less than 9 per cent of the pairs wanted to meet up again, compared to 18 per cent when participants spoke about the most popular topic – travel. A clue as to why would-be lovers might want to avoid chatting about movies comes from additional data from the study. At the start of the evening we asked everyone to indicate their favourite types of film. The results revealed that men and women have very different tastes. For instance, 49 per cent of men liked action films compared to just 18 per cent of women, whilst 29 per cent of women liked musicals compared to only 4 per cent of men. Whenever I walked past the table having conversations about films, all I heard was people arguing. In contrast, the conversations about travel tended to revolve around great holidays and dream destinations, and that makes people feel good and so appear more attractive to one another.

The data also revealed other surprises. Although men are traditionally seen as being shallow and judging women very quickly, our findings suggested that women were making up their minds much quicker than men, with 45 per cent of women's decisions being made in under thirty seconds, compared to just 22 per cent of men's decisions. Men clearly have only a few seconds to impress a woman, emphasizing the importance of their opening comments.

To uncover the best type of chat-up lines, we compared the conversations of participants rated as very desirable by their dates with those seen as especially undesirable. Failed Casanovas tended to employ old chestnuts like 'Do you come here often?' or struggled to impress with comments such as 'I have a PhD in computing', and 'My friend is a helicopter pilot'. Those more skilled in seduction encouraged their dates to talk about themselves in an unusual, fun, quirky way. The most memorable lines from the top-rated man and woman in the study illustrate the point. The top-rated male's best line was: 'If you were on *Stars In Their Eyes*, who would you be?' whilst the top-rated female asked: 'If you were a pizza topping, what would you be?'

Why should these latter types of lines be so successful? The answer lies in a rather unusual experiment involving drinking straws and funny voices.

In 2004, psychologists Arthur Aron (of the 1974 bridge study) and Barbara Fraley, from the State University of New York at Stony Brook, randomly paired strangers and had them carry out one of two sets of slightly strange behaviours.[46] In one condition, one of the strangers was blindfolded, and the other was asked to hold a drinking straw between their teeth (which made their voice sound funny). The two then carried out a series of tasks designed to make them laugh. The blindfolded person had to learn a series of dance steps by listening to instructions read by their straw-holding partner. In another example of laboratory-based hilarity, they were asked to act out their favourite television commercial using a made-up language. The other, more straight-faced, condition did not involve drinking straws. Here, the dance steps were learnt without the blindfold and silly voice, and the commercials were acted in the English language. Everyone was then asked to complete a questionnaire about how much fun they had had, with the results confirming that the blindfold, drinking straw, and silly language had resulted in significantly more hilarity. Then came the crunch question. All of the participants were asked to draw two overlapping circles to indicate the degree of closeness that they felt with their partner. The results revealed that people who had had the shared humour experience felt significantly closer to their partner, and found them more attractive.

The successful chat-up lines in our study were the speed dating equivalent of putting a straw in your mouth to make your voice sound silly. In doing so, they elicited a shared funny experience that promoted a sense of closeness and attraction.

'Minimalist seeks woman': The psychology of personal ads

Imagine that you are going to write a personal ad. What choice of words do you think would prove most successful and attract the largest number of replies? This was exactly the question tackled in another aspect of our journey into the science of seduction.

We asked everyone involved in the speed dating experiment to write a short personal ad containing about twenty words. We then showed these to over a hundred men and women, and asked them to indicate which ads they would be most likely to answer. The results provided important clues to a hitherto unexplored aspect of ads.

Previous work into personals had examined the type of person most frequently sought by men and women.[47] The results haven't yielded too many big surprises. Men tend to look for women who are physically attractive, understanding, and athletic. In contrast, women are searching for someone who is understanding, humorous, and emotionally healthy. I decided to take a different tack.

I looked through the ads we had received and noticed something odd. There was a large variation in the number of words that each person used to describe themselves, compared with the number of words used to describe the person that they were looking for. Which type of ad would attract the greatest number of replies – the one that describes you in greater detail or the one that describes the person you are looking for?

To find out, I counted the number of words that each person had used to describe themselves, and the type of person that they were looking for. I then used these two numbers to derive a 'self versus other' percentage. At one extreme were the 'it's all about you' people who obtained a near 0 per cent score by saying very little about themselves, and instead focusing almost entirely on their wish list:

> Brunette, 27, looking for someone kind, romantic, spontaneous, caring, and who is willing to take a risk. We can always tell them we met in the supermarket!

In the middle of the range were the 'it's about the two of us' people who split the wording more evenly, describing both themselves and their potential partner, and obtained a 50 per cent score:

> Laid-back guy, good sense of humour, into sport, travel, lethal coffee, eating out, seeks creative, funny, sunny, happy, charismatic girl to while away long summer nights.

Then at the other end of the spectrum were the 'it's all about me' people who obtained a 100 per cent score by focusing almost entirely on themselves:

> Bright, fun, gym-loving, non-smoker, singer-songwriter, into detective novels, funny films, American comedy shows, and long walks on sunny beaches.

I next looked at the relationship between the score assigned to each ad, and the number of people indicating that they would reply to it. The results were revealing. Only a small number of people indicated that they would reply to 'it's all about you' ads. The 'it's all about me' ads fared a little better, but still didn't attract many replies. A balance between the two extremes turned out to be the winning formula. The results showed that a 70 per cent 'this is me' versus 30 per cent 'this is what I am looking for' balance attracted the greatest number of replies. It seems that if you devote more than 70 per cent of the ad to describing yourself, you look self-centred. Less than 70 per cent and you look suspicious.

Our two top ads fitted the pattern, and contained the rough 70:30 split. Just over 45 per cent of men said that they would reply to the winning female ad:

> Genuine, attractive, outgoing professional female, good sense of humour. Enjoys keeping fit, socialising, music, and travel. Would like to meet like-minded, good-natured guy to share quality times.

Similarly, almost 60 per cent of women indicated that they would be attracted to the top male ad:

> Male, good sense of humour, adventurous,
> athletic, enjoys cooking, comedy, culture, film,
> seeks sporty, fun female for chats and possible
> romantic relationship.

Our study also provided another top tip for those wishing to write winning ads. We asked our group of one hundred people to indicate which ads they thought members of the opposite sex would be likely to answer. The results showed a remarkable difference between the sexes.

First, let's have a look at the ads written by men. We compared the percentage of women who said that they would reply to each ad with the percentage of men who *thought* that women would reply. So, one of the ads read:

> Tall, slim, athletic, fashionable male with a
> good sense of humour looking for slim to aver-
> age girl with good sense of humour interested
> in cars, music, clothes and cuddles.

About 11 per cent of women said that they would reply to the ad. Interestingly, men said that they thought 15 per cent of women would reply – a remarkably accurate prediction. Another ad noted:

> Tall, energetic male with his head in the clouds
> and feet on the ground. Would like to meet a
> woman who is fun, positive and isn't afraid of
> a challenge.

This time, 39 per cent of women ticked the 'yes' box. Again, men's predictions were remarkably accurate, predicting that 32 per cent of women would answer this ad. And so it went on. For ad after ad, men were able to accurately predict which ads women would find attractive and which they would avoid like the plague. Overall, men's predictions were, on average, 90 per cent correct.

A very different story emerged when it came to women predicting men's behaviour. Look at the following ad, written by a female:

> Cute and quirky professional with a passion for good food, wine and company looking for the proverbial tall, dark and handsome with cracking wit and fit body.

In reality, this didn't appeal to men. Only 5 per cent said that they would reply to the ad. Yet women were convinced that it would act as a man-magnet, predicting that about 44 per cent would reply. How about:

> Relaxed, upbeat, friendly woman who enjoys relaxing, laughing, exploring the world, and wants to dance the night away!

Once again, women thought that this would attract the majority of men, but were wrong. Only about 22 per cent of men indicated that they would respond.

The pattern repeated itself across the ads. Women simply had very little idea about what actually attracted men. So why are women so inaccurate? Perhaps the unsolicited 'questionnaire graffiti' suggesting that women thought men were only interested in the physical attributes of women is a clue. Time and again, we came across comments such as 'they are just interested in one thing', and 'are only interested by two things'. Our research suggests that perhaps men are not quite so shallow. Regardless, the implication for women using personals is simple: if you want to attract lots of potential beaus, get a guy to write your ad.

5

The scientific search for the world's funniest joke: Explorations into the psychology of humour

How a surreal international quest revealed why men and women find different jokes funny, the link between laughter and longevity, what makes professional comedians tick, and whether weasels really are the world's funniest animal.

In the 1970s, the cult comedy show *Monty Python's Flying Circus* created a sketch based entirely around the idea of finding the world's funniest joke. Set in the 1940s, a man named Ernest Scribbler thinks of the joke, writes it down, and promptly dies laughing. The joke turns out to be so funny that it kills anyone who reads it. Eventually, the British military realize that it could be used as a lethal weapon, and have a team of people translate the joke into German. Each person translates just one word at a time, in order not to be affected by the joke. The joke is then read out to German forces, and is so funny that they are unable to fight because they are laughing so much. The sketch ends with footage from a special session of the Geneva Convention, in which delegates vote to ban the use of joke warfare.

In a strange example of life imitating art, in 2001 I headed a team carrying out a year-long scientific search for the world's funniest joke. Instead of exploring the potential military application of jokes, we wanted to take a scientific look at laughter.

In addition to finding the joke that had maximum mass appeal, my Pythonesque project resulted in a string of surreal experiences involving American humorist Dave Barry, a giant chicken suit, Hollywood actor Robin Williams, and over 500 jokes ending with the punchline 'there's a weasel chomping on my privates'.

More importantly, the project also provided considerable insights into many of the questions facing modern-day humour researchers. Do men and women laugh at different types of jokes? Do people from different countries find the same things funny? Does our sense of humour change over time? And if you are going to tell a joke involving an animal, are you better off making the main protagonist a duck, a horse, a cow, or a weasel?

Why did the chicken cross the road?

In June 2001, I was contacted by the same august scientific body that had commissioned my study into financial astrology, the British Association for the Advancement of Science (BAAS). The BAAS were eager to create a project that would act as a centrepiece for a year-long national celebration of science, and wanted a large-scale experiment that would attract public attention. Would I be interested in creating it, and, if so, what would I choose to investigate?

After a few 'close, but no cigar' moments, I happened to see a rerun of the *Monty Python* sketch involving Ernest Scribbler, and I started to think about the possibility of *really* searching for the world's funniest joke. I knew that there would be a firm scientific underpinning for the project, because some of the world's greatest thinkers, including Freud, Plato and Aristotle, had written extensively about humour. In fact, the German philosopher Ludwig Wittgenstein was so taken with the topic that he once stated that a serious work in philosophy could be written entirely of jokes. I then discovered that whenever I mentioned my idea to people, it provoked a serious amount of discussion. Some queried whether there really was such a thing as the world's funniest joke. Others thought that it was impossible to analyse humour scientifically. Almost everyone was kind enough to share their favourite joke with me. The rare mix of good science and popular appeal meant that the idea felt right.

I presented the BAAS with my plans for an international Internet-based project called 'LaughLab'. I would set up a website that had two sections. In one part, people could input their favourite joke and submit it to an archive. In the second section, people could answer a few simple questions about themselves (such as their sex, age, and nationality), and then rate how funny they found various jokes randomly selected from the archive. During the course of the year, we would slowly build a huge collection of jokes and ratings from all around the globe, and be able to discover scientifically what makes different groups of people laugh, and which joke made the whole world smile. Everyone at the BAAS nodded, and LaughLab got the green light.

The success of the project hinged on being able to persuade thousands of people worldwide to come online and participate. To help spread the word, the BAAS and I launched

LaughLab by staging an eye-catching photograph based around perhaps the most famous (and, as we would go on to prove scientifically, least funny) joke in the world: 'Why did the chicken cross the road? To get to the other side.' It was September 2001, and I found myself standing in the middle of a road dressed in a white laboratory coat and holding a clipboard. Next to me was a student dressed in a giant chicken suit. Several national newspaper photographers were lined up in front of us, snapping away, and I can still vividly remember one of them looking up and shouting: 'Can the guy playing the scientist move to the left?' I shouted back, 'I *am* a scientist,' and then looked sheepishly at the giant chicken standing next to me. It was the type of surreal experience that was to occur all too frequently throughout the next twelve months.

The launch was successful, and LaughLab made its way into newspapers and magazines across the globe. Within a few hours of opening the website for business, we received over 500 jokes and 10,000 ratings. Then we hit a major problem. Many of the jokes were a tad rude. Actually, I am understating the issue. They were absolutely filthy. One especially memorable submission involved two nuns, a large bunch of bananas, an elephant, and Yoko Ono. We couldn't allow these submissions into the archive because we had no control over who would visit the site to rate the jokes. With a backlog of over 300 jokes from the first day alone, we clearly needed someone to work full-time to vet them. My research assistant, Emma Greening, came to the rescue. Every day for the next few months, Emma carefully looked at every joke and excluded those that were not suitable for family viewing. She was often frustrated by seeing the same jokes again and again (the joke 'What is brown and sticky?' 'A stick', was submitted over 300 times), but, on the upside, now owns one of the largest collections of dirty jokes in the world.

Participants were asked to rate each joke on a five-point scale ranging from 'not very funny' to 'very funny'. To simplify our analyses, we combined the '4' and '5' ratings to make a general 'yes, that is quite a funny joke' category. We could then order the jokes on the basis of the percentage of responses that fell into this category. If the joke really wasn't very good, then it might have only 1 per cent or 2 per cent of people assigning it a '4' or '5' rating. In contrast, the real rib-ticklers would have a much higher percentage of top ratings. At the end of the first week, we reviewed some of the leading submissions. Most of the material was pretty poor, and so tended to obtain low percentages. Even the top jokes fell well short of the 50 per cent mark. Around 25–35 per cent of participants found the following jokes funny, and so they came towards the top of the list:

A teacher decided to take her bad mood out on her class of children and so said, 'Can everyone who thinks they're stupid, stand up!' After a few seconds, just one child slowly stood up. The teacher turned to the child and said, 'Do you think you're stupid?'

'No . . .' replied the child, '. . . but I hate to see you standing there all by yourself.'

Did you hear about the man who was proud when he completed a jigsaw within thirty minutes, because it said 'five to six years' on the box?

Texan: Where are you from?

Harvard graduate: I come from a place where we do not
    end our sentences with prepositions.

Texan: Okay – where are you from, jackass?

An idiot was walking along a river, when he spied another
idiot on the other side of the river. The first idiot yelled to
the second idiot: 'How do I get to the other side?' The
second idiot responded immediately: 'You're already *on*
the other side!'

The top jokes had one thing in common – they create a
sense of superiority in the reader. The feeling arises because
the person in the joke appears stupid (like the man with the
jigsaw), misunderstands an obvious situation (like the idiots
on the riverbank), pricks the pomposity of another (like the
Texan answering the Harvard graduate), or makes someone in a position of power look foolish (like the teacher and
the child). These findings provided some empirical support
for the adage concerning the difference between comedy and
tragedy: 'If *you* fall down an open manhole, that's comedy.
But if *I* fall down the same hole . . .'

We were not the first to notice that people laugh when
they feel superior to others. The theory dates back to around
400 BC, and was described by the Greek scholar Plato in his
famous text *The Republic*. Proponents of this 'superiority'
theory believe that the origin of laughter lies in the baring of
teeth akin to 'the roar of triumph in an ancient jungle duel'.
Because of these animalistic and primitive associations, Plato
was not a fan of laughter. He thought that it was wrong to
laugh at the misfortune of others, and that hearty laughter

involved a loss of control that resulted in people appearing to be less than fully human. In fact, the father of modern-day philosophy was so concerned about the potential moral damage that could be caused by laughter that he advised citizens to limit their attendance at comedies, and never to appear in this lowest form of the dramatic arts.

Plato's view was echoed in the later writings of fellow Greek thinker Aristotle. Unfortunately, we have only indirect references to Aristotle's thoughts on the subject, because his original treatise on the topic has become lost in the mists of time (and is the manuscript that lies at the heart of Umberto Eco's mystery *The Name of the Rose*). Aristotle apparently argued that many successful clowns and comedians make us laugh by eliciting a sense of superiority. It is easy to find support for the theory. In the Middle Ages, dwarves and hunchbacks caused much merriment. In Victorian times, people laughed at the mentally ill in psychiatric institutions, and at those with physical abnormalities in freak shows. There is also the 1976 study showing that when the public were asked to list adjectives describing comedians, they tended to offer the words 'fat', 'deformed', and 'stupid'.[1]

The superiority theory is also used to poke fun at entire groups of people. The English traditionally tell jokes about the Irish, the Americans like to laugh at the Polish, the Canadians pick on the Newfies, the French on the Belgians, and the Germans on the Ostfriedlanders.[2] In each case it is about one group of people trying to make themselves feel good at the expense of another.

In 1934, Professor Wolff and his colleagues published the first experimental study of the superiority theory.[3] The researchers asked groups of Jews and Gentiles to rate how funny they found various jokes. To ensure that the presentation of the jokes was as controlled as possible, the researchers printed them along strips of cloth 140 feet long and 4 inches wide, and then passed the strips behind an aperture in the laboratory wall at a constant speed, ensuring that participants saw each joke one word at a time. When participants saw a star symbol that had been printed at the end of each strip, they were asked to shout out how funny they had found the joke on a scale between −2 (very annoying) and +4 (very humorous). As predicted by Aristotle and Plato, the Gentiles tended to laugh more at jokes disparaging Jews, whilst the Jews preferred the jokes that put down the Gentiles. Another part of the experiment explored whether a 'control' group – the Scots – would prove equally amusing to both Jews and Gentiles. They presented participants with a series of anti-Scottish jokes, such as the classic 'Why are Scotsmen so good at golf? The fewer times they hit the ball, the less it will wear out', and were surprised to find that the Gentiles found them significantly funnier than the Jews. The experimenters initially wondered whether this might have been because Gentiles have a better sense of humour than Jews, but then realized that the anti-Scottish jokes were a bad choice of control. Both the Jews and the Scots are often portrayed as 'stingy' in jokes, and this had caused the Jews to sympathize with the Scots, and find the anti-Scottish jokes unfunny. Taking part in this groundbreaking study apparently wasn't easy on people. Some participants complained that they had heard many of the jokes before, with one man noting that he would rather be subjected to electric shocks than additional one-liners.

Modern researchers have worked hard to overcome these

problems, and their findings have helped expand and refine the superiority theory.

We now know that the more superior a joke makes a person feel, the harder they laugh. Most people do not find a disabled person slipping on a banana skin funny, but replace the disabled person with a traffic warden and suddenly everyone is slapping their thighs. This simple idea explains why so many jokes attack those in power, such as politicians (thus David Letterman's famous quip: 'The traffic was so bad I had to squeeze through spaces that were narrower than President Clinton's definition of sex'), and judges and lawyers ('What do you call a lawyer with an IQ of 10? A lawyer. What do you call a lawyer with an IQ of 15? Your honour'). People in positions of power often do not see the funny side of these, and treat them as a real threat to their authority. Hitler was sufficiently concerned about the potential use of humour, that he set up special 'Third Reich joke courts' that punished people for acts of inappropriate humour, including naming their dogs 'Adolf'.[4]

Some research suggests that jokes like these can have surprisingly serious consequences. In 1997, psychologist Gregory Maio from Cardiff University of Wales and his colleagues looked at the effect that reading superiority jokes had on people's perception of those who were the butt of the jokes.[5] The study was carried out in Canada, and so centred around the group who was frequently portrayed as stupid by Canadians, namely Newfoundlanders (or 'Newfies'). Before the experiment, participants were randomly assigned to one of two groups. The people in each group were asked to read one of two sets of jokes into a tape recorder, supposedly to help determine the qualities that make a voice sound funny or unfunny. Those in one group read jokes that did not involve laughing at Newfies (such as Seinfeld material), whilst the other read Newfie put-down humour (such as the classic thigh-slapper: 'A Newfie friend of mine heard that every minute a woman gives birth to a baby. He thinks she should be stopped'). Afterwards, everyone was asked to indicate their thoughts about the personality traits of Newfoundlanders. Those who had just read out the Newfie jokes rated Newfoundlanders as significantly more inept, foolish, dim-witted, and slow, than those who had delivered the Seinfeld material.

Just as worrying, other work has revealed that superiority jokes have a surprisingly dramatic effect on how people see themselves.[6] Professor Jens Förster, from the International University Bremen in Germany, recently tested the intelligence of eighty women of varying hair colour. Half of them were asked to read jokes in which blondes appeared stupid. Then all participants took an intelligence test. The blonde women who had read the jokes obtained significantly lower scores on the IQ test than their blonde counterparts in the control condition, suggesting that jokes have the power to affect people's confidence and behaviour, and so actually create a world in which the stereotypes depicted in the jokes become a reality.

Very early in LaughLab, we could see the superiority theory appearing by virtue of the age-old battle of the sexes. The following joke was rated as being funny by 25 per cent of women, but just 10 per cent of men:

A husband stepped on one of those penny scales that tell you your fortune and weight and dropped in a coin.

'Listen to this,' he said to his wife, showing her a small white card. 'It says I'm energetic, bright, resourceful and a great person.' 'Yeah,' his wife nodded, 'and it has your weight wrong, too.'

One obvious possibility for the difference in ratings between the sexes is that the butt of the joke is a man, and so appeals more to women. However, that is not the only possible interpretation of the result. It could be, for example, that women generally find jokes funnier than men. A year-long study of 1,200 examples of laughing in everyday conversation revealed that 71 per cent of women laugh when a man tells a joke, but only 39 per cent of men laugh when a woman tells a joke.[7] To help try to tease apart these competing interpretations, we studied the LaughLab archive to find jokes that put down women, such as:

A man driving on a highway is pulled over by a police officer. The officer asks: 'Did you know your wife and children fell out of your car a mile back?' A smile creeps onto the man's face and he exclaims: 'Thank God! I thought I was going deaf!'

On average, 15 per cent of women rated jokes putting down women as funny, compared to 50 per cent of men. The points awarded to these jokes revealed that the superiority theory did explain the differences between what makes men and women laugh. But this is not to say that there are no differences between the sexes when it comes to humour and jokes. Research suggests that men tell a lot more jokes than women. In one study, over 200 college students were asked to record all of the jokes that they heard during a one-week period, and make a note of the joke-teller's sex. The group reported 604 jokes, with 60 per cent of them coming from men rather than women.[8] This difference in joke-telling has been observed in many different countries, and is present even when children first start to tell jokes to other another.[9]

Some scholars believe that the difference can be attributed to the fact that women avoid jokes because they may be of a sexual nature, or involve acts of aggression ('What do you call a monkey in a minefield? A BABOOM!'). Others think that the difference has its roots in the link between laughter, jokes, and status. People with high social status tend to tell more jokes than those lower down the pecking order. Traditionally, women have had a lower social status than men, and thus may have learnt to laugh at jokes, rather than tell them. Interestingly, the only exception to this status – joke-telling relationship concerns self-disparaging humour, with people who have low social status telling more self-disparaging jokes than those with high status. In line with this idea, researchers examining the amount of self-disparaging humour produced by male and female professional comedians found that 12 per cent of male scripts contained self-disparaging humour, compared with 63 per cent of female scripts.[10]

We took our first in-depth look at our data three months into the project. The project's technical guru, Jed Everitt, downloaded the 10,000 jokes, and the ratings from the 100,000 people who had been kind enough to tell us how funny they found each of the rib-ticklers. The top joke at that early stage had been rated as funny by 46 per cent of participants. It had been submitted by Geoff Anandappa, from Blackpool in the north-west of England, and involved the famous fictional detective Sherlock Holmes and his long-suffering sidekick, Dr Watson:

> Sherlock Holmes and Dr Watson were going camping. They pitched their tent under the stars and went to sleep. Sometime in the middle of the night Holmes woke Watson up and said: 'Watson, look up at the stars, and tell me what you see.'
>
> Watson replied: 'I see millions and millions of stars.'
>
> Holmes said: 'And what do you deduce from that?'
>
> Watson replied: 'Well, if there are millions of stars, and if even a few of those have planets, it's quite likely there are some planets like earth out there. And if there are a few planets like earth out there, there might also be life.'
>
> And Holmes said: 'Watson, you idiot, it means that somebody stole our tent.'

It is a classic example of two-tiered superiority theory. We laugh at Watson for missing the absence of the tent, and also at the pompous way in which Holmes delivered the news to Watson.

Two thousand years ago, the ancient Greek philosopher Plato speculated that the sense of superiority plays a key role in the creation of humour. Our findings suggested that not only was he right, but that the animalistic release of a victorious roar at other people's misfortune is still alive and well in the twenty-first century.

'A cigar may just be a cigar, but a joke
is never just a joke'

Although the initial stages of the experiment had been a huge success, we still wanted more people to visit our virtual laboratory. Because of this, we announced our initial findings to the media. After the success of our 'Why did the chicken cross the road?' photograph, we staged a second striking photograph involving an actor dressed as a clown lying on the actual couch used by the famous psychologist Sigmund Freud. Why Freud? Well, he was fascinated by humour, and in 1905 produced his classic treatise on the topic, *Jokes and their Relation to the Unconscious.*

Freud's basic model of the mind revolved around the notion that we all have sexual and aggressive thoughts, but that society does not allow us to express these ideas openly. As a result, they become repressed deep in our unconscious and emerge only via the odd slip of the tongue (the 'Freudian slip'), in dreams, and in certain forms of psychoanalysis.

According to Freud, jokes act as a kind of psychological release valve that helps prevent the pressure of repression from becoming too great – in other words, a way of dealing with whatever it is that causes us to feel anxious. The simple act of telling a joke, or laughing at someone else's joke, reveals a great deal about the unconscious, causing him once to quip: 'A cigar is sometimes just a cigar, but a joke is never *just* a joke.' Given that, in Freudian terms, a cigar is often seen as symbolizing a penis, I have always found Freud's choice of image for his famous line interesting.

There is a great deal of debate about Freud's contribution

to the psychology of humour, with one group of academics noting that '. . . it would be exceedingly difficult to find a person of at least average intelligence who knows less about humour than did Freud'.[11] There again, this from a group of researchers who described the superiority theory of humour using the following paragraph:

> Let S believe J is a joke in which A seems to S victorious and/or B appears the butt. Then the more positive S's attitude towards A and/or towards the 'behaviour' of A, and/or the more negative S's attitude towards B and/or towards the 'behaviour' of B, the greater the magnitude of amusement S experiences with respect to J.

The Freud Museum is based in the North London house where the great mind doctor worked during the final phase of his life. The building contains a wonderful collection of books and artefacts, and, of course, Freud's famous couch. This five-foot-long chaise longue was apparently given to him by a grateful patient in the 1890s. During a typical therapy session, a patient would recline on the couch, and Freud would sit in a large armchair. He devised various techniques to reach the activities of the unconscious. Sometimes he would ask a patient to talk about their dreams. At other times Freud would say a certain word and have the patient respond with the first word that came into his or her head. The couch has come to symbolize Freud's approach to understanding the human mind, and so provided the perfect backdrop for the second LaughLab photograph. The BAAS contacted the Museum, and were delighted when the director granted us special permission to have a clown recline on this most famous of couches.

On a cold December morning in 2001, the LaughLab team (plus clown) arrived at the Museum and were shown into Freud's office. It is an impressive room. One wall is lined with bookshelves containing Freud's extensive collection of books and manuscripts. Scattered around the room are a remarkable collection of Egyptian, Greek and Roman antiquities. The couch sits in one corner of the room next to Freud's large leather armchair.

The photographers arrived and we took up our positions. Our clown carefully reclined on the couch, and I picked up a clipboard and took my place in the armchair. Sitting in the chair that had once been occupied by the world's most famous psychologist, and being greeted from his couch by a man wearing a huge bright-blue wig, an overdrawn grease-paint grimace, and massive red shoes, proved to be another surreal LaughLab moment. The photographers liked the set-up, and merrily snapped away. To help induce a sense of realism into the pictures, one of them asked if I could I conduct an informal therapy session with the clown. Although not a Freudian psychologist, I was happy to try. I asked my 'patient' what the problem was, and the quick-thinking clown said that he was upset because no one took him seriously.

Although Freud claimed to be a scientist, many of his ideas are completely untesticle. Even so, many of the jokes submitted to LaughLab certainly supported Freud's ideas. Time and again, we would get jokes about the stresses and strains of loveless marriage, inadequate sexual performance, and, of course, death:

I've been in love with the same woman for forty years. If my wife finds out she'll kill me.

A patient says to his psychiatrist: 'Last night I made a Freudian slip. I was having dinner with my mother-in-law and wanted to say: "Could you please pass the butter." But instead I said: "You silly cow, you have completely ruined my life."'

A guy goes to the hospital for a check-up. After weeks of tests, a doctor comes to see him and says that he has some good news and some bad news.

'What's the bad news?' asks the man.

'I am afraid we think you have a very rare and incurable disease,' says the doctor.

'Oh, my God, that's terrible,' says the man. 'What's the good news?'

'Well,' replies the doctor, 'we are going to name it after you.'

Some of the submissions allowed us to explore Freud's theories. Given that older people tend to be especially anxious about the effects of ageing, would they find gags about memory loss, and the like, funnier than younger people? Freud would argue that this should be the case, but would our data support this? We carefully sifted through the joke archive, and selected several jokes that centred around the difficulties associated with getting old, like the following:

An elderly couple had dinner at another couple's house, and after eating, the wives left the table and went into the kitchen.

The two elderly gentlemen were talking, and one said: 'Last night, we went out to a new restaurant, and it was really great. I would recommend it very highly.'

The other man said: 'What was the name of the restaurant?'

The first man thought and thought and finally said: 'What is the name of that flower you give to someone you love? You know . . . the one that is red and has thorns.'

'Do you mean a rose?'

'Yes,' the man said. He then turned towards the kitchen and yelled: 'Rose, what's the name of that restaurant we went to last night?'

and:

A man in his late sixties suspects that his wife is going deaf, so he decides to test her hearing. He stands on the opposite side of the living room from her and asks: 'Can you hear me?' No answer. He moves halfway across the room towards her and asks: 'Can you hear me now?' No answer. He moves and stands right beside her and says: 'Can you hear me now?' She replies: 'For the third time, yes!'

The results were as predicted by Freud. Younger people didn't like these types of jokes. On average, about 20 per cent of people under the age of 30 found each joke funny, versus 50 per cent of people in the 'sixty or over' category. The message is clear – we laugh at the aspects of life that cause us the greatest sense of anxiety.

We also inadvertently conducted a second experiment testing this idea. Emma Greening, our joke-vetting expert, had done a grand job of keeping the rude material away from the website. However, she did allow one joke through by mistake:

A guy goes to his priest and says 'I feel terrible. I am a doctor and I have slept with some of my patients.' The priest looks concerned, and then tries to make the man feel better by saying, 'You aren't the first doctor to sleep with their patients and you won't be the last. Perhaps you shouldn't feel so guilty.'

'You don't understand,' says the man. 'I'm a vet.'

It is a classic Freudian joke, and revolves around one of the most basic societal taboos – sex with animals. Interestingly, it obtained a very high score, with about 55 per cent of people finding it funny. Draw your own conclusions from the fact that men found it funnier than women, and people from Denmark found it funniest of all.

The humour of the hemispheres

Scientists are not known for their sense of humour. However, since we were conducting an experiment, we thought it appropriate to approach some of Britain's best-known scientists and science writers, and ask them to submit their favourite jokes to LaughLab. They all proved obliging, and we ended up receiving material from some of the UK's top thinkers, including the Director of the Royal Institution, Baroness Susan Greenfield; planetary scientist and principal investigator of the ill-fated Beagle 2 Mars lander project, Professor Colin Pillinger; evolutionary biologist Professor Steve Jones; and best-selling science author, Dr Simon Singh.

The joke that went on to win the 'best joke submitted by a well-known scientist or science writer' category was submitted by Nobel laureate and professor of chemistry, Sir Harry Kroto. Kroto is best known for being part of the team that discovered a new form of carbon known as C60 Buckminsterfullerene, and not quite so well known for describing himself as adhering to four 'religions': humanism, atheism, amnesty-internationalism, and humourism. It may have been this last interest that gave him the edge over his fellow scientists, with his winning joke being that old chestnut involving two men and a dog:

> A man walking down the street sees another man with a very big dog. The man says: 'Does your dog bite?'
>
> The other man replies: 'No, my dog doesn't bite.'
>
> The first man then pats the dog, has his hand bitten off, and shouts: 'I thought you said your dog didn't bite.'
>
> The other man replies: 'That's not my dog.'

Overall, the jokes submitted by scientists did not fare especially well. In fact, they came in the bottom third of all jokes submitted, and even Sir Harry Kroto's winning entry beat only 45 per cent of other jokes.[12]

We also examined another source of humour – computers. LaughLab attracted lots of jokes about this topic. However, it also contained a few jokes actually written by a computer.

A few years ago, Dr Graham Ritchie and Dr Kim Binsted created a computer program that could produce jokes.[13] We were keen to discover whether computers were funnier than humans, and so entered several of the computer's best jokes into LaughLab. The majority of them received some of the lowest ratings in the archive. One example of computer comedy, however, was surprisingly successful, and beat about

250 human jokes: 'What kind of murderer has fibre? A cereal killer.'

It's an example of the most basic form of joke – the simple pun. The most popular theory about why we find this sort of joke funny revolves around the concept of 'incongruity'. The idea is that we laugh at things that surprise us because they seem out of place. It's funny when clowns wear outrageously large shoes (especially when they are not performing), when people have especially big noses, or when politicians tell the truth. In the same way, many jokes are funny because they involve ideas that run against our expectations. A bear walks into a bar. Animals and plants talk. But there is more to this theory than simple forms of incongruity. In many jokes, there is an incongruity between the set-up and the punchline. For example:

> Two fish in a tank. One turns to the other and says: 'Do you know how to drive this?'

The set-up line leads us to think about two fish in a fish tank. But the punchline surprises us – why should the fish be able to drive a fish tank? Then, a split second later, we realize that the word 'tank' has two meanings, and that the fish are actually in an army tank. Scientists refer to this as the 'incongruity-resolution' theory. We resolve the incongruity caused by the punchline, and the accompanying feeling of sudden surprise makes us laugh.

The LaughLab team decided to discover what was happening in people's brains when they laughed at these types of jokes. To help, I contacted neuroscientist Dr Adrian Owen at Cambridge University. I chose Adrian for two reasons. First, he is one of the leading brain imagers in the world. Second, the two of us studied psychology together at college, devised and performed in the 'Captain Fearless' magic show during our summer breaks, and, despite it all, remain good friends. Adrian teamed up with colleague Professor Steve Williams from the Institute of Psychiatry and used a technique known as functional magnetic resonance imaging (or fMRI for short) to examine what was happening inside people's brains when they laughed at some of the best puns from the project.

Brain-scanning is used to study all sorts of psychological phenomena. One of my favourite experiments was carried out by Professor Gert Holstege at the University of Groningen, and examined how women fake orgasms.[14] The study involved women with their head inside a scanner, while their partner stimulated them manually to achieve a real orgasm. The women were also asked to fake an orgasm. Comparing the two scans revealed that faking was associated with certain parts of the brain, providing a very expensive way of knowing whether an orgasm was genuine. The researchers also discovered that many couples were put off because their feet were cold. When they gave the couples socks to wear, about 80 per cent of the couples were able to achieve orgasm compared with 50 per cent in the 'no sock' condition.

Our scans were far more straightforward to obtain, but no less surreal. The work involved carefully placing people's heads inside a million-pound scanner, and asking them to read some of the top-rated puns. The results revealed that the left side of the brain plays a key role in setting up the initial context for the joke ('There are two fish in a tank'), whilst a small area in the right hemisphere provides the creative skills necessary to realize that the situation can be seen in a completely different, and often surreal, way ('One fish turns to the other and says: Do you know how to drive this?'). One of the resulting brain scans is shown below. This scan shows two areas in the left hemisphere being activated after being shown some of the set-up lines from LaughLab jokes.

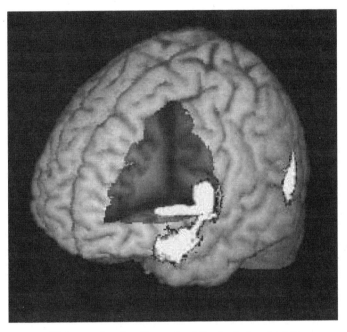

A 3D scan showing the parts of the brain involved in finding jokes funny.

This work supported other research showing that people who have experienced damage to the right hemisphere are less able to understand jokes, and so don't see the funny side of life.[15] Have a look at the following set-up line and then the three possible punchlines, and see if you can choose the correct one.

A man went up to a lady in a crowded square. 'Excuse me,' he said, 'do you happen to have seen a policeman anywhere around here?'

'I'm sorry,' the woman answered, 'but I haven't seen one for ages.'

Potential punchlines:

a) 'Oh, OK, can you give me your watch and necklace, then.'
b) 'Oh, OK, it's just that I've been looking for one for half an hour.'
c) 'Baseball is my favourite sport.'

The first punchline is obviously correct. The second one makes sense, but isn't funny. And the third doesn't make sense, *and* isn't funny. People with damage to the right side of the brain tended to choose the third punchline far more often than people who did not have any brain damage. It seems that these people know that the end of the joke should be surprising, but have no way of knowing that one of the punchlines could be reinterpreted to make sense. Interestingly, they still find films of slapstick comedians funny – they haven't lost their sense of humour, but rather have lost the ability to work out why certain incongruities are funny, and others aren't. Some of the researchers conducting this work summed up the situation by noting that: 'While the left hemisphere might appreciate some of Groucho's puns, and the right hemisphere might be entertained by the antics of Harpo, only the two hemispheres united can appreciate a whole Marx Brothers routine.' As journalist Tad Friend dryly noted in his *New Yorker* article on LaughLab, neither hemisphere seems to find Chico funny.[16]

Weasels, and the comedy K

In January 2002, Emma Greening walked into my office and said, 'I just don't get it, we are receiving a joke every minute, and they all end with the same punchline: "There's a weasel chomping on my privates".' We were five months into the project and, unbeknown to us, the internationally syndicated American humorist Dave Barry had just devoted an entire column to our work in the *International Herald Tribune*.[17] In a previous column, Barry claimed that any sentence can be made much funnier by the insertion of the word 'weasel'.[18] In his column concerned with LaughLab, Barry repeated his theory, and urged his readers to submit jokes to our experiment that ended with the punchline, 'There's a weasel chomping on my privates'. In addition, he asked people to assign any weasel joke a maximum 5 points whenever it was chosen from the archive. Within just a few days we had received over 1,500 'weasel chomping' jokes.

Barry is not the only humorist to develop a theory about which words, and sounds, make people laugh. The results from one of the mini-experiments that we conducted during LaughLab, supported the most widely cited of these theories: the mysterious comedy potential of the letter 'K'.

Early in the experiment, we received the following submission:

> There were two cows in a field. One said: 'Moo.' The other one said: 'I was going to say that!'

We decided to use the joke as a basis for a little experiment. We re-entered the joke into archive several times, using a different animal and noise. We had two tigers going 'Gruurrr', two birds going 'Cheep', two mice going 'Eeek', two dogs going 'Woof', and so on. At the end of the study, we examined what effect the different animals had had on how funny people found the joke. In third place came the original cow joke; second were two cats going 'Meow', but the winning animal-noise joke was:

> Two ducks were sitting in a pond. One of the ducks said: 'Quack.' The other duck said: 'I was going to say that!'

Interestingly, the 'K' sound (as in the 'hard C'), associated with both the words 'quack' and 'duck', has long been seen by comedians and comedy writers as being especially funny.

The idea of the 'comedy K' has certainly made it into popular culture. In the *Star Trek: The Next Generation* episode 'The Outrageous Okona', a comedian refers to the idea when attempting to explain humour to the android Data. There was also an episode of *The Simpsons* in which Krusty the Clown (note the 'K's) visits a faith healer because he has paralysed his vocal chords trying to cram too many 'comedy Ks' into his routines. After being healed, Krusty exclaims that he is overjoyed to get his comedy Ks back, celebrates by shouting out 'King Kong', 'cold-cock', 'Kato Kaelin', and kisses the faith healer as a sign of gratitude.

Why should the 'K' sound produce such pleasure? It may be due to a rather odd psychological phenomenon known as 'facial feedback'. People smile when they feel happy. However, there is some evidence to suggest that the mechanism also works in reverse: that is, that people feel happy simply because they have smiled.

In 1988, Professor Fritz Strack and his colleagues had people judge how funny they found Gary Larson *Far Side* cartoons under one of two conditions.[19] One group of participants were asked to hold a pencil between their teeth, but to ensure that it did not touch their lips. Without the people realizing it, this forced the lower part of their faces into a smile. The other group were asked to support the end of the pencil with just their lips, and not their teeth. This forced their faces into a frown. The results revealed that people actually experience the emotion associated with their expressions. Those who had their faces forced into a smile felt happier, and found the *Far Side* cartoons much funnier, than those who were forced to frown.

Interestingly, many words featuring the 'K' sound force the face into a smile (think 'duck' and 'quack'), and may account for why we associate the sound with happiness. Regardless of whether this explains the 'comedy K' effect, the explanation certainly plays a key role in another aspect of humour – the contagious nature of laughter.

In 1991, American psychologists Verlin Hinsz and Judith Tomhave, from North Dakota State University, visited various shopping centres to examine smiling.[20] One of the team smiled at a random selection of people whilst another experimenter, secreted inside a fake food stand, carefully observed whether the person reciprocated with a smile. After hours of smiling and observing, they discovered that half of people responded to the experimenter's smile with another smile. Their results caused them to suggest that the old saying, 'Smile, and the whole world smiles with you', should be altered to the more scientifically accurate: 'Smile, and *half* the world smiles with you'.

The ability automatically, and unconsciously, to mimic the facial expressions of those around us plays a vitally important role in group survival, cohesion and bonding. By copying the expressions of others, we quickly feel how they feel, and so find it much easier to empathize with their situation and communicate with them. One person in the group smiles, and others automatically copy their facial expression and cheer up. Another feels sad or afraid or panicked and, again, the emotion they display can spread from person to person. This, combined with the results from the pencil experiment, explains why laughter is contagious. When people see or hear another person laugh, they are far more likely to copy the behaviour, start to laugh themselves, and therefore actually find the situation funny. This is the reason why so many television comedy programmes carry laughter tracks, and why nineteenth-century theatre producers would hire a 'professional' audience member (known as the *rieur*) who had an especially infectious laugh, in order to encourage the whole audience to giggle and guffaw.

Although this contagion usually has limited effects, sometimes it can get out of hand, and an otherwise inexplicable epidemic of laughter can sweep through thousands of people. In January 1962, three teenage girls attending a missionary-run boarding school in Tanzania started laughing.[21] Their hilarity quickly spread to 95 of the 159 pupils at the school, and by March the school was forced to close. It is reported that the attacks of laughing lasted from minutes to hours, and, although debilitating, did not result in any fatalities. The school reopened in May, but was forced to close again within weeks when another sixty pupils were struck down with the 'laughter plague'. The closure created its own problems, with several of the girls returning to their hometown of Nshamba, promptly causing over 200 of the 10,000 residents of the town to descend into uncontrollable giggling. It is not known whether their teacher had a 'K' in his or her name.

Comedians on the couch

After a few months we had received more than 25,000 jokes, about a million ratings, and a large amount of international publicity. It was around this time that I was contacted by John Zaritsky, an Oscar-winning Canadian documentary-maker, asking whether I would like to help make a film based on LaughLab and examining humour across the world. I quickly agreed, and together we travelled the world looking at what makes different parts of the globe guffaw, titter, and groan.

As part of the film, John invited me to Los Angeles to road-test some of the material that was obtaining high ratings. I carefully searched through the database and identified two types of jokes – those that were found especially funny by the British and those that appealed to Americans. In June 2002, I found myself standing in the wings of the Ice House, a comedy club in Pasadena, California. The compère, a sassy young woman named Debi Gutierrez, was standing on stage explaining what was about to happen. She described the LaughLab project, and said that I would come on stage and deliver some of the jokes that had proved most popular with the British, whilst she would deliver the gags that had received high ratings from the Americans. A few moments later, and I was on. It was another of those surreal moments. Debi opened with a classic:

> *Woman to male pharmacist:* 'Do you have that Viagra
>   drug?'
> *Pharmacist:* 'Yes.'
> *Woman:* 'Can you get it over the counter?'
> *Pharmacist:* 'Only if I take two of them.'

Debi managed to mess up the punchline, and the joke fell flat. Not even a titter. Then it was my turn. I decided to open with a 'Doctor, Doctor' joke that had proved very popular with the British people visiting our virtual laboratory:

> A guy goes to the doctor and has a checkup. At the end
> of the examination, he turns to the doctor and asks how
> long he has to live. The doctor replies, 'Ten.' The guy
> looks confused, and says, 'Ten what? Years? Months?
> Weeks?' The doctor replies, 'Nine, Eight, Seven . . .'

Again, it was a tumbleweed moment. You could have heard a pin drop. Or a duck drop, if it's funnier. After a few more jokes obtained exactly the same negative response, Debi finally produced the only laugh of the session with an improvised opening of a non-existent joke: 'Two faggots and a midget walk into a bar . . .'

According to our data, around a third of the audience should have found the jokes funny. In reality, that figure was embarrassingly close to zero. So what went wrong? It was a case of horses for courses. The people voting in our experiment represented a wide cross-section of the public, whereas the people at the comedy club were into a certain type of comedy – bold, brash, offensive, and aggressive. When it comes to comedy, there is no magic bullet – no single joke that everyone will find really funny. It is all a question of matching the joke to the person, and we had missed by a mile. It was a point that was to come up repeatedly when we announced our winning joke at the end of the experiment.

Although being on stage at the Ice House was no fun, waiting backstage with the other performers proved to be more interesting. Professional comedians are, if you will excuse the pun, a funny group of people. They have chosen to make their living in a difficult and highly stressful way. They have to walk out on stage, night after night, and make a group of complete strangers laugh out loud. No matter how they feel, or what is happening in their own lives, they have to be funny. In view of this, a small number of psychologists have been interested in analysing their minds.

Woody Allen once remarked: '. . . most of the time I don't have much fun, the rest of the time I don't have fun at all'. But how much truth is there in the popular stereotype of the sad clown? It is certainly easy to think of high-profile examples, including British comedian Spike Milligan (who suffered with manic depression throughout his life), and American performers Lenny Bruce and John Belushi (both of whom are believed to have committed suicide).

In 1975, American psychiatrist Samuel Janus published a groundbreaking paper on the psychology of comedians. Janus was keen to investigate the truth behind the popular notion of the sad clown, and so interviewed fifty-five well-known professional comedians about their lives.[22] Janus gathered together some of the top names in comedy, working only with those who were earning at least a six-figure salary and were known nationally. The results revealed that the vast majority of the group possessed above average intelligence (with few achieving the classification of 'genius'), 80 per cent had sought psychotherapy at some point in their lives, and nearly all were extremely anxious about the possibility of their star status fading away. This latter finding caused Janus to conclude that, 'Several were able to enjoy life and reap the benefits of their fame and fortune, but they were in a very small minority.' His report also shows the problems of working with highly successful, but angst-ridden, professional comedians. Although almost all performed well on the tests of intelligence, Janus noted that, 'The problem was not one of getting them to respond, it was one of continuously allaying their anxiety, and reassuring them that they were indeed doing well.' Also, when asked about their experience with psychotherapy, Janus describes how several of his participants said that the therapist had asked them to '. . . lie down on the couch and tell me everything you know', shortly followed by the comment, 'And now he's doing my act in Philadelphia.'

Pretend the world is funny and forever

Janus's image of 'comedian as sad clown' was not supported by work published in 1981 by the psychoanalysts Seymour and Rhoda Fisher from the State University of New York at Syracuse.[23] The Fishers conducted an extensive investigation into more than forty well-known comedians and clowns, including Sid Caesar, Jackie Mason, and Blinko the Clown, and published their findings in a wonderful book entitled *Pretend the World is Funny and Forever*.

As part of the work, they administered a classic Freudian test known as the Rorschach ink blot test, in which their participants were asked to look at an ambiguous ink blot and say what it reminded them of. The test has been used extensively in research and even features in a well-known Freudian joke:

> A man goes to a psychoanalyst. The analyst gets out a stack of cards containing ink blots, shows them to the man one at a time, and asks him to say what the ink blots remind him of. The man looks at the first ink blot and says, 'Sex'. Then he looks at the second ink blot and says, 'Sex' again. In fact, he goes through the whole stack of images, saying the word 'Sex' in response to every one. The psychiatrist looks concerned and says, 'I don't wish to worry you, but you seem to have sex on your mind.' The man looks surprised, and answers, 'I can't believe you just said that – you are the one with all the dirty pictures.'

Most of the tests took place in restaurants and circus dressing rooms, with the Fishers describing how they would often find the test difficult to administer because of constant interruptions by members of the public and fellow performers.

Contrary to both the popular conception of the 'sad clown', and Janus's previous results, the Fishers found little evidence of psychopathology. Given the stressful nature of working as a professional comedian, the Fishers were surprised to discover that their interviewees appeared remarkably resilient and well adjusted.

Another aspect of the Fishers' work examined the childhood experiences of the comedians and clowns. They noted that the majority of their interviewees started young, and were often considered the 'class clown'. In line with the 'superiority' theory, many related instances in which they used humour to get one up on their teachers. One performer recalled how the teacher had asked him to go to the blackboard and spell 'petroleum'. He promptly walked up to the front of the class, picked up a piece of chalk, and wrote the word 'oil'. Professional comedians tended to come from relatively low-income families, and may have experienced an unhappy childhood, and thus their performances may represent an attempt to compensate by gaining the affection of an audience. There is considerable anecdotal evidence to support the idea. Woody Allen once said that the 'need to be accepted' was one of his primary motivations for being funny, Jack Benny didn't enjoy a holiday in Cuba because no one recognized him, and W. C. Fields once explained that he liked making people laugh because: '. . . at least for a short moment, they love me'.

A third element of the Fishers' research examined the psychological qualities associated with being funny. Several performers admitted that they were intensely curious about people and behaviour, reporting how they would endlessly watch others go about their lives until they found some small idiosyncrasy that could form the basis for a new joke or routine. The Fishers noted that there were many parallels between comedians and social scientists. They argued that both groups are constantly on the prowl for novel perspectives on human behaviour, with the only major difference being that in one instance such insights make people laugh, whilst in the other they form the basis for academic papers. Having spent my career reading the latter publications, I would venture that this alleged distinction fails to separate the two groups clearly.

The Fishers also set out to examine the relationship between comedy and anxiety. When presented with ink blots, people often see one image and then realize that the blot can be seen in a different way. After carefully analysing the types of images that the comedians reported seeing in the random ink blots, the Fishers concluded that their participants often produced a unique class of 'nice monster' imagery, wherein threatening figures would be transformed into something far more pleasant. A 'dragon with flames shooting out of his mouth' would become a noble and misunderstood figure, and a 'dirty hyena' would change into a lovely, cuddly pet. The Fishers interpreted this as evidence of comedians and clowns being unconsciously motivated to use humour as a way of coping with events that might otherwise cause considerable distress.

They are not the only ones to question the notion that comedy is, deep down, associated with sadness and psychopathology. James Rotton from Florida International University examined a copy of Hoffman's *Entertainment Personalities of the Past* for the years of birth and death of well-known comedians, and compared it with a control group of non-comedic entertainers born in the same years. (Rotton reports restricting his list to male comedians, after discovering that the ages of many female comedians were unreliable as they didn't match other biographic sources, suggesting a possible chronopsychology of comedy.)[24] Describing his findings in a paper entitled *Trait Humor and Longevity: Do Comics Have the Last Laugh?*, Rotton argued that comedians were no more likely to die young than other entertainers. A subsequent analysis of the cause of death of comedians (gleaned by examining the obituaries of performers published in *Time* and *Newsweek* between 1980 and 1989) revealed no evidence of an excess of heart attacks, cancers, pneumonia, accidents, or suicides. In short, there is no evidence to suggest that the obvious stresses and strains associated with having to be funny night after night leads to an early demise.

Rotton's findings are in line with other work suggesting that being able to laugh at life reduces anxiety and that, if anything, comedy may actually be good for your health. In the thirteenth century, surgeon Henri de Mondeville speculated that laughter might promote recovery from illness, writing: 'The surgeon must forbid anger, hatred, and sadness in the patient, and remind him that the body grows fat from joy and thin from sadness.' A few hundred years later, William Shakespeare echoed the same sentiment, noting: 'Frame your mind to mirth and merriment, which bars a thousand harms and lengthens life.'

Recent work has supported a link between laughter, coping with stress, and psychological and physical well-being. According to this work, people who spontaneously use humour to cope with stress have especially healthy immune systems, are 40 per cent less likely to suffer a heart attack and strokes, experience less pain during dental surgery, and live 4.5 years longer than most.[25]

In 1990 researchers discovered that watching a video of Bill Cosby performing his stand-up routine results in enhanced production of salivary immunoglobulin A – a chemical that plays a key role in preventing upper respiratory tract infection (apparently these beneficial effects were significantly reduced when participants listened to Mel Brooks and Carl Reiner's classic '2000-Year-Old-Man' routine).[26] This is not the only work exploring the bodily effects of laughter. In 2005, Michael Miller, and his colleagues from the University of Maryland, studied the relationship between finding the world funny and the inner lining of blood vessels. When such vessels expand they increase blood flow around the body, and promote cardiovascular well-being. Participants were shown scenes from films that were likely to make them either feel anxious (such as the opening thirty minutes of *Saving Private Ryan*), or laugh (such as the 'orgasm' scene from *When Harry Met Sally*). Overall, participants' blood flow dropped by around 35 per cent after watching the stress-inducing films, but rose by 22 per cent following the more humorous material. On the basis of the results, the researchers recommend that people laugh for at least fifteen minutes each day.

In a similar vein, James Rotton examined the effects that watching different kinds of videos had on hospital patients recovering from orthopaedic surgery.[27] One group of patients were asked to select funny films from a list including *Bananas*, *Naked Gun*, and *The Producers*. Another group were denied access to any material that might induce a smile, and were instead asked to select movies from a 'serious' list, including titles such as *Brigadoon*, *Casablanca*, and *Dr No*. The experimenters secretly monitored the quantity of major pain-relievers that the patients consumed via a self-controlled pump. Those watching funny films used just over 60 per cent less pain-relieving drugs than those looking at the serious movies. In an interesting twist to the experiment, the researchers also included another group of patients who were not allowed to select which comedy films to watch, but instead were given the movies selected by others. This group administered significantly more drugs than either of the other groups, scientifically proving that there is nothing more painful than watching a comedy that doesn't make you laugh.

Finally, a team of researchers asked participants to reflect on their mortality by constructing a mock will, completing their own death certificate (including an estimate of their date and cause of death), and writing the eulogy for their own funeral.[28] Researchers discovered that those who had exhibited a prior tendency to laugh at the absurdities of life did not find the tasks as stressful as more gloomy participants. Exactly the same effect has emerged in more realistic settings. Bereavement counsellors interviewed people six months after they had lost a spouse, and found that those who could laugh about the loss were more able than others to come to terms with the situation and move on with their lives.[29] However, as one of the jokes submitted to LaughLab illustrates, it is possible to take this idea too far:

A man dies and his wife telephones her local newspaper, and says, 'I would like to print the following obituary: Bernie is dead.'

The man at the newspaper pauses, and says, 'Actually, for the same price you could print six words.'

The woman replies, 'Oh, okay. Can I go with: Bernie is dead. Toyota for sale.'

Heard the one about the religious fundamentalists?

Given the physical and psychological benefits of laughter, it perhaps isn't surprising that some scientists have investigated the characteristics of people who do, and do not, see the funny side of life. Some of the most intriguing work in the area has been carried out by psychologist Vassilis Saroglou, from the Université Catholique de Louvain in Belgium, and investigated the relationship between laughter and religious fundamentalism.

Saroglou argues that there is a natural incompatibility between religious fundamentalism and humour.[30] The creation and appreciation of humour requires a sense of playfulness, an enjoyment of incongruity ('Two fish in a tank . . .'), and a high tolerance of uncertainty. Humour also frequently involves mixing elements that don't go together, threatens authority, and contains sexually explicit material. In addition, the act of laughter involves a loss of self-control and self-discipline. All of these elements, argues Saroglou, are the antithesis of religious fundamentalism, with research showing that those who subscribe to it tend to value serious activities over playfulness, certainty over uncertainty, sense over nonsense, self-mastery over impulsiveness, authority over chaos, and mental rigidity over flexibility. Saroglou supports his argument by noting that many scholars have written about the deep mistrust that exists between humour and religion, citing evidence from various religious texts. When discussing examples of Biblical humour, Saroglou notes:

> One might wonder, for instance, why Christ didn't simply say 'blessed are you that weep now, for you shall laugh' (Lk 6:21), but went on to add 'woe to you that laugh now, for you shall mourn and weep' (Lk 6:25).

Additional evidence comes from a monastic rule regarding priests playing with children:

If a brother willingly laughs and plays with children . . .
he will be warned three times; if he does not stop, he will
be corrected with the most severe punishment.

Eager to test his hypothesis empirically, Saroglou carried out
a rather unusual experiment.[31] In one part of the study, participants completed a questionnaire to measure their level of
religious fundamentalism, rating the degree to which they
endorsed various ideas; including whether a particular set
of teachings contain fundamental truths, that these truths
are opposed by evil forces, and that they must be followed
via a set of well-defined historical practices. In another part
of the experiment, the participants were shown a set of
twenty-four pictures depicting a variety of frustrating everyday situations, and asked to note down how they would
react. After completing the picture task, the researchers rated
the degree of humour in participants' responses. For instance,
one of the cards showed someone falling to the ground in
front of two friends, with one of the friends asking, 'Did
you hurt yourself?' A straight-faced response to this card
might be something like, 'No, I'm fine', whilst 'I don't know,
I haven't reached the ground yet' would constitute a far more
humorous approach. In line with his prediction, Saroglou
found a strong relationship between religious fundamentalism and humour, with fundamentalists producing far more
serious-sounding answers than others.

As is almost always the case with research showing a relationship between two factors, it is difficult to separate cause
and effect. Perhaps having a poor sense of humour leads to a
fundamentalist religious belief. Or maybe, as hypothesized by
Saroglou, being a fundamentalist prevents people from seeing

the funny side of life. To help disentangle these two possibilities, Saroglou carried out a second study.[32] This time he split participants into three groups. Two of the groups were shown very different film footage. One group saw funny footage from famous French comedy shows. A second group saw religiously oriented footage, including a documentary film about a pilgrimage to Lourdes, scenes from *Jesus of Montreal*, and a discussion between a journalist and a monk about spiritual values. A third group didn't see any film footage at all, and so acted as a control. Participants were then asked to complete the same humour production task as before. Overall, the people who had seen the humorous footage produced over twice as many funny responses as the control group, and those who had seen the religious footage trailed in third place. The clever design of the experiment helped Saroglou differentiate correlation from causation, and suggests that exposure to religious material actually prevents people from using humour to help ease the stressful effects of everyday hassles.

The relationship between humour production and religious fundamentalism didn't stop people submitting jokes about various deities to the LaughLab experiment. Many of them played on the incongruous idea of an omnipotent, caring God behaving out of character:

A shipwreck survivor washes up on the beach of an island and is surrounded by a group of warriors.

'I'm done for,' the man cries in despair.

'No, you are not,' comes a booming voice from the heavens. 'Listen carefully, and do exactly as I say. Grab a spear and push it through the heart of the warrior chief.'

The man does what he is told, turn to the heavens, and asks, 'Now what?'

The booming voice replies, '*Now* you are done for.'

Robin Williams, Spike Milligan, and the 'woof woof' joke

In a 1939 edition of the *New York Times Magazine*, Canadian humorist Stephen Leacock remarked, 'Let me hear the jokes of a nation and I will tell you what the people are like, how they are getting along, and what is going to happen to them.' Our data allowed us to take a scientific look at Leacock's idea by examining national differences in humour. We were not the first academics to tackle the topic.

In chapter 1 I described the groundbreaking work into astrology and personality carried out by the prolific British psychologist Professor Hans Eysenck. During the Second World War, Eysenck developed an interest in the psychology of humour, and carried out an unusual survey into British, American, and German magazine cartoons.[33] Getting hold of the material proved problematic. Large shipments of American magazines had been lost through sinkings in the Atlantic, much of the potential British material had been destroyed in the bombings of the British Museum, and the choice of German material was restricted to magazines published before the outbreak of hostilities. Despite the problems, Eysenck soldiered on, and eventually managed to find seventy-five cartoons drawn from various magazines, including *Punch* in Britain, the *New Yorker* in America, and *Berliner Illustrierte* in Germany.

Eysenck then translated the German captions into English, and showed all of the cartoons to groups of English people. Everyone was first asked to rate how funny they found the cartoons. Eysenck discovered that the cartoons from all three countries obtained roughly the same funniness ratings. Next, participants were asked to try to guess whether each cartoon originated in Britain, America, or Germany, allowing the experimenters to analyse the points awarded to the cartoons as a function of where the participants *thought* that they were from. The cartoons that people thought were from Germany received much lower ratings than those believed to be from America and Britain. Further analyses revealed additional evidence of national stereotypes. When Eysenck analysed the cartoons that people thought were German, he found an over-representation of various negative elements, including fat women, badly dressed girls, and old-fashioned furniture.

In a second part of his study, Eysenck had British, American, and German volunteers (actually refugees who had fled their homeland because of the war), rate the same set of jokes and limericks. His results revealed that, overall, Americans found the material funnier than people from the other two countries, but that there were few differences in the types of jokes that people from different countries found funny.

Our LaughLab data broadly supported Eysenck's findings. There were large differences in how funny the jokes were to people from different countries. Canadians laughed least at the jokes. This could be seen in one of two ways. Given that the jokes were not very good, it could be argued that Canadians have a discerning sense of humour. Alternatively, they may simply not have much of a sense of humour at all, and so not find anything funny. Germans found the jokes funnier than people from any other country. The validity of this finding was questioned within the pages of almost every non-German newspaper and magazine that reported our findings. One British newspaper told the top-rated German joke ('Why is television called a medium? Because it is neither rare nor well done') to a spokesperson at the German embassy in London. Apparently, he laughed so much he dropped the telephone, and cut off the call. Few other differences emerged. On the whole, people from one country found the same jokes funny and unfunny. Comedian and musician Victor Borge once described humour as the shortest distance between two people. If he is correct, then knowing that diverse groups of people laugh at the same jokes might help bring such groups closer together.

By the end of the project we had received 40,000 jokes, and had them rated by more than 350,000 people from seventy countries. We were awarded a Guinness World Record for conducting one of the largest experiments in history, and made the cover story of the *New Yorker*. The top joke, as voted by Americans, was as follows:

> At the parade, the Colonel noticed something unusual going on and asked the Major: 'Major Barry, what the devil's wrong with Sergeant Jones' platoon? They seem to be all twitching and jumping about.'
>
> 'Well sir,' says Major Barry after a moment of observation, 'there seems to be a weasel chomping on his privates.'

Dave Barry had been successful, and managed to make the top American joke weasel-oriented. He had, thank goodness, less influence over the votes cast by those outside of America.

We carefully went through the huge archive and found our top joke. It had been rated as funny by 55 per cent of the people who had taken part in the experiment:

> Two hunters are out in the woods when one of them collapses. He doesn't seem to be breathing and his eyes are glazed. The other guy whips out his phone and calls the emergency services. He gasps, 'My friend is dead! What can I do?' The operator says, 'Calm down. I can help. First, let's make sure he's dead.' There is silence, then a shot is heard. Back on the phone, the guy says 'OK, now what?'

The joke had just beaten the initial front-runner, concerning Holmes, Watson and the strange case of the missing tent, to first place. We contacted Geoff Anandappa, the man who had submitted the Holmes and Watson joke, and broke the bad news. Geoff was gracious in defeat, noting: 'I can't believe I got knocked out in the final round! I could've been a contender . . . I want a rematch, and this time I'm going to fight dirty. Did you hear the one about the actress and the bishop?'

The database told us that the winning joke had been submitted by a psychiatrist from Manchester in Britain named Gurpal Gosall. We contacted Gurpal and he explained how he sometimes told the joke to cheer up his patients, noting that: '. . . it makes people feel better, because it reminds them that there is always someone out there who is doing something more stupid than themselves'.

The BAAS and I announced our findings at the third, and final, press conference. For the last time, we hired a chicken suit, and had one of my lucky doctorate students dress up (see overleaf).

A giant chicken reveals the world's funniest joke

Our winning joke was printed on a huge banner, and unveiled to the waiting press. Comedian Terry Jones, one of the Python team responsible for the sketch involving the world's funniest joke, was asked by the media for his opinion.[34] He thought the joke was quite funny but perhaps rather obvious. Another journalist interviewed Hollywood star Robin Williams about our winning entry. Like Jones, Williams thought that it was okay, but went on to explain that the world's funniest joke is probably absolutely filthy and therefore not the sort of thing you would tell in polite company.

The year-long search for the world's funniest joke concluded. Did we really manage to find it? In fact, I don't believe it exists. If our research into humour tells us anything, it is that people find different things funny. Women laugh at jokes in which men look stupid. The elderly laugh at jokes involving memory loss and hearing difficulties. Those who are powerless laugh at those in power. There is no one joke that will make everyone guffaw. Our brains just don't work like that. In many ways, I believe that we uncovered the world's blandest joke – the gag that makes everyone smile but very few laugh out loud. But, as with so many quests, the journey was far more important than the destination. Along the way we looked at what makes us laugh, how laughter can make you live longer, how humour should unite different nations, and we discovered the world's funniest comedy animal.

Five years after the study, I received a telephone call from my friend and brain-scanning scientist, Adrian Owen. He explained that he had just seen a documentary film about Spike Milligan, comedian and co-founder of the Goons, and that the programme contained a very early version of our winning joke. The documentary (the title of which, *I Told You I Was Ill*, was based on Spike's epitaph) contained a brief clip from a 1951 BBC programme called *London Entertains* with the following early Goon sketch:

Michael Bentine: I just came in and found him lying on
the carpet there.
Peter Sellers: Oh, is he dead?
Michael Bentine: I think so.
Peter Sellers: Hadn't you better make sure?
Michael Bentine: All right. Just a minute.
(*Sound of two gunshots.*)
Michael Bentine: He's dead.

It is highly unusual to be able to track down the source of a joke, because their origins tend to become lost in the mists of time. Spike Milligan had died in 2002, but with the help of the documentary-makers, I contacted his daughter Sile, and she confirmed that it was highly likely that her father would have written the material. We announced that we believed we had identified the author of the world's funniest joke, and LaughLab made headlines again.

During the ensuing interviews, several journalists asked me a question that frequently occurs whenever I speak about LaughLab: What was my favourite joke from the thousands that flooded in through the year? I always give the same reply:

A dog goes into a telegraph office, takes a blank form and writes:

'Woof woof woof. Woof, woof. Woof. Woof woof, woof.'

The clerk examines the paper and politely tells the dog: 'There are only nine words here. You could send another "Woof" for the same price.'

The dog looks confused and replies, 'But that would make no sense at all.'

I don't really know why I like it. It just makes me laugh.

6

Sinner or saint? The psychology of when we help, and when we hinder

Why fake leg braces have been worn to measure altruism around the globe, how dropping envelopes across America revealed whether Catholics are more giving than most, the secret psychology used to create caring communities, and the mystery of the vanishing teaspoons.

In the early 1930s, Stanford psychologist Richard LaPiere spent several months driving up and down the West Coast of America with one of his Chinese students, and the student's wife.[1] The couple had been born and raised in China, and had only recently moved to the States. To them, LaPiere appeared to be a genial professor who had kindly found the time to show them around. In reality, they were unsuspecting guinea pigs in a secret experiment that LaPiere was conducting throughout the trip.

The idea for the experiment had occurred to LaPiere when the couple had first arrived at the university, and he had taken them to the main hotel in town. A relative rarity in 1930s America, Chinese people were frequently subjected to a considerable amount of blatant prejudice. According to LaPiere, he approached the hotel with a sense of trepidation because it was '. . . noted for its narrow and bigoted attitudes towards Orientals'.

LaPiere went to the reception with his two friends, and nervously asked whether they had any rooms available. To LaPiere's surprise, the clerk didn't display the prejudice for which his establishment had gained a considerable reputation, and instead quickly found them some suitable accommodation. Curious about the discrepancy between what he had heard about the hotel, and his experience with the receptionist, LaPiere later telephoned the hotel and asked whether they would have a room available for 'an important Chinese gentleman'. He was told in no uncertain terms that the hotel would not provide accommodation.

LaPiere was struck by the discrepancy between how people *said* that they would behave, and how they *actually* acted. But he realized that his experience at the hotel could just have been atypical. To investigate the issue properly he would need to repeat the same scenario with a far larger number of hotels and restaurants, and that is when he hit upon the idea of taking his two Chinese colleagues on an experimental road trip across America.

The journey involved driving 10,000 miles, and visiting 66 hotels and 184 restaurants. At each hotel and eatery, LaPiere had his student ask about the possibility of accommodation or food. LaPiere then secretly noted whether their request was successful. The results from this initial part of the study replicated his earlier experience. His two companions received pleasant and helpful service almost everywhere they went, leading LaPiere to conclude that:

> . . . the 'attitude' of the American people, as reflected in the behavior of those who are for pecuniary reasons presumably most sensitive to the antipathies of their white clientele, is anything but negative towards the Chinese.

Six months later, LaPiere conducted the second part of the study. He sent questionnaires to each of the hotels and restaurants that they had visited, asking, 'Will you accept members of the Chinese race as guests in your establishment?' To help hide the exact purpose of the study, this question was one of many, with others enquiring about whether each establishment welcomed Germans, French, Armenians, and Jews. The results made disturbing reading. Over 90 per cent of respondents ticked the 'no, Chinese people are not welcome here' box, with almost all of the remaining 10 per cent going for the 'uncertain' option. LaPiere received only one 'yes' response. This reply came from a hotel that LaPiere and his students had visited a few months before. The owner had added a short note to the questionnaire saying that the reason she would welcome people from China was because she had recently enjoyed a nice visit from a Chinese man and his sweet wife.

In LaPiere's study, people said that they would behave in a way that was in keeping with the societal norms of the day, but actually behaved quite differently. In more recent studies, researchers have obtained copious evidence for the same effect, with people claiming that they are not racist (in keeping with modern societal norms), but then behaving in a prejudiced way.

It all adds up to one simple point. Asking people to rate how nice they are is unlikely to yield a genuine insight into anything other than their ability to deceive themselves and others. Because of people's reluctance, or inability, to report accurately whether they are nice or nasty, many researchers interested in these topics have done exactly what LaPiere did. They have stopped asking people to tick the 'saint' or 'sinner' box, taken off their lab coats, put on trenchcoats, and carried out secret studies in the real world.

Vanishing gloves, attaché cases, and female van drivers

For the past twenty-five years, Professor John Trinkaus, from the City University of New York, has dedicated his academic life to the scientific observation of ordinary folk going about their everyday business. Publishing his findings in almost one hundred academic papers, Trinkaus has investigated a huge range of topics. He has visited railway stations and noted the colour of sports shoes worn by men and women (79 per cent of male sports shoe wearers chose white, versus just 34 per cent of female wearers[2]), counted the number of times television weather forecasters *said* that their predictions had been accurate with the number of times they were *actually* correct (only 49 per cent of the allegedly accurate predictions were right[3]), travelled to inner-city estates to document the decreasing numbers of people wearing baseball caps with their peaks turned to the back (it is dropping at a rate of 10 per cent per year[4]), and plotted verbal trends in providing an affirmative response by counting the number of times that interviewees on television chat shows used the word 'yes' when answering questions (of the 419 questions analysed, 'yes' was used 53 times, 'exactly' 117 times, and 'absolutely' 249 times[5]).

Trinkaus's investigations into sports shoes, weather forecasts, baseball cap wearing, and the use of the 'yes' word, are all brilliantly obscure. Some of his other research, however, has serious implications, not least his work into the surprising predictability of human nature.[6] Trinkaus asked hundreds of his students to think of any odd number between 10 and 50, and found the majority chose 37. When asked to name any even number between 50 and 100, most said 68. Trinkaus then took this aspect of his research into the real world, asking a hundred people who owned attaché cases with number locks to tell him the combination.[7] He discovered that almost 75 per cent of owners had not changed the factory settings on their cases, and that they could be opened with the numbers 0-0-0. (In his book, *Surely You're Joking, Mr Feynman*,[8] physicist Richard Feynman describes how he used the same type of predictability to gain access to top-secret documents when he was working on the development of the atomic bomb at an American army base in Los Alamos. On one occasion he opened a colleague's safe by trying various combinations he thought a physicist might use – the winning numbers were 27-18-28, after the mathematical constant, $e = 2.71828$. Another time Feynman discovered that no one had bothered to change the generic factory settings to one of the largest safes on the base, which could have been opened by an unskilled thief within minutes.)

My favourite piece of Trinkaus research is described in his little-known paper, 'Gloves as vanishing personal "stuff": An informal look'.[9] Here Trinkaus begins by noting how his personal belongings, including single socks, umbrellas, and one glove from each pair, often seem to disappear. He then goes on to explain that he has managed to overcome the problem with umbrellas by purchasing several inexpensive ones from a street vendor (who, according to his observations, charged 50 per cent more for them on a rainy than a sunny day), but is reluctant to apply the same approach to gloves.

Eager to get to the bottom of the vanishing gloves mystery, Trinkaus monitored his missing mittens during a ten-year period, carefully noting whether the vanishing glove belonged to his left or right hand. The results revealed that over three times as many left- as right-hand gloves went missing. This caused him to speculate that he might be removing his right-hand glove first, pushing it down into his pocket, then removing his left-hand glove, and pushing it on top of the glove already there. If this was true, then his left-hand glove would be nearer the top of the pocket, and more likely to fall out during the course of the day.

Trinkaus's work into vanishing gloves has inspired other researchers to investigate similar topics. In 2005, researchers Megan Lim, Margaret Hellard, and Campbell Aitken, from the MacFarlane Burnet Institute for Medical Research in Melbourne, conducted an experiment to discover why teaspoons in communal kitchens disappear with annoying regularity (or, as they phrase it in their scientific paper on the topic, to answer the age-old question 'Where have all the bloody teaspoons gone?')[10] The team secretly marked seventy teaspoons, placed each of them in one of eight communal kitchens at their Institute, and tracked the movement of the spoons over a five-month period. Eighty per cent of the teaspoons went missing during this time, with half of them disappearing within the first eighty-one days. Additional questionnaire data revealed that 36 per cent of people said they had stolen a teaspoon at some point in their lives, with 18 per cent admitting to such a theft in the last twelve months. This latter result argues against the notion that the vanishing spoons are being sucked into another dimension, and instead supports a more down-to-earth explanation: people steal them.

The researchers also note that the Institute's level of disappearing teaspoons, multiplied by the entire Melbourne workforce, suggests that eighteen million teaspoons go missing each year in Melbourne alone; if these spoons were laid end to end, they would stretch around the coastline of Mozambique. Unlike the Trinkaus research into vanishing gloves, other researchers have started to replicate the 'disappearing teaspoons' study across the globe. In one of the most recent pieces of follow-up work, French academics reported that 1,800 teaspoons went missing over a six-month period in a large cafeteria.[11]

The stealing of teaspoons brings us indirectly to Trinkaus's work into dishonesty and antisocial behaviour. As we shall see later in this chapter, many of the other researchers interested in these types of behaviour are drawn to serious acts of stealing or selfishness. Trinkaus has managed to develop his own unique approach, focusing on relatively small-scale social transgressions, such as people taking more than ten items through the express line in a supermarket, or parking in restricted areas. His findings have revealed a surprising insight into just how common these occurrences are, how they can be used to chart the moral decline of society, and the relationship between them and female van drivers.

In 1993, Trinkaus and his team of researchers visited a large supermarket in the north-east of America, and secretly observed customers on seventy-five different occasions for a period of fifteen minutes each.[12] They carefully counted how many people took more than ten items through the 'ten items or less' line. To help ensure the scientific validity of their study, they observed shoppers at various times of the day over the course of several weeks, and recorded people's behaviour only if more than two other lanes were open (and they therefore had the option of taking their goods through a correct checkout). The results revealed that around 85 per cent of shoppers in the express lines were breaking the rules by having more than ten items in their basket. In 2002, Trinkaus repeated the same experiment at the same super- market, and discovered that the percentage of deceptive shoppers had risen to 93 per cent.[13] Projected forward, these figures suggest that by 2011, no one in the express line will have ten items or less in their basket.

Trinkaus also noticed a new form of dubious behaviour that had developed since his 1993 study. Several shoppers in the 'ten items or less' line placed their items on the conveyor belt in groups of ten, and then told the cashier that they would pay for each group separately. One shopper managed to get twenty-nine items through the express line using this sneaky approach. As soon as Trinkaus spotted this new form of deceptive behaviour, he realized that it could be used as a potential way of identifying the types of people most likely to transgress societal norms. In line with his observational approach to research, Trinkaus asked his team to follow these people into the supermarket car park, and make a note of their gender and the type of vehicle they owned. The result: about 80 per cent of the transgressors were female van drivers.

This is not the first time that Trinkaus has uncovered evidence suggesting that female van drivers are especially likely to indulge in antisocial behaviour. In 1999, he counted and classified the number of motorists speeding near a school, and noted that 96 per cent of female van drivers exceeded the speed limit, compared to just 86 per cent of male van drivers.[14] In the same year, he had also been counting the number of motorists who failed to come to a complete stop at T-junctions with stop signs.[15] In total, 94 per cent of motorists failed to comply with the sign, versus 99 per cent of female van drivers. In 2001, he spent thirty-two hours logging 200 instances in which motorists had failed to keep a boxed intersection clear, and found that 40 per cent of incidents involved, yes, you guessed it, women driving vans.[16] A year later, he counted instances of people parking their vehicles in a prohibited fire zone at a shopping centre.[17] Again, women van drivers were the least compliant, accounting for about 35 per cent of all violations.

Trinkaus has put forward two explanations to account for this aspect of his data. First, he has speculated that 'women van drivers are inadvertently carrying over from the workplace the now "in" concept of empowerment'. According to this approach, women are still getting used to their new-found power in society, and may have developed an unconscious need to outdo behaviours previously associated with men, such as breaking speed limits, parking in restricted areas, and ignoring traffic signs. Alternatively, notes Trinkaus, these drivers may be ahead of the game in terms of the moral decline of society, and acting as an omen of things to come.

Testing the honesty of a nation

In 1997, the *US News and World Report* conducted a poll in which they asked Americans who was 'somewhat likely' to go to heaven.[18] Bill Clinton didn't do too badly, with 52 per cent of respondents thinking that he would be welcome at the pearly gates. Princess Diana fared a little better with 60 per cent of the vote, and in second place came Mother Teresa with 79 per cent. But who won the poll, scoring a massive 87 per cent? Most people placed *themselves* top of the heavenly A-list. Do we really live in communities and countries populated almost entirely by potential saints?

A few years ago, I was asked by a television programme called *World In Action* to help devise several tests that would examine the honesty of the nation. Rather than adopt Trinkaus's approach of observing people carrying out relatively minor transgressions, we decided to be a little more proactive and focus on serious acts of selfishness. Many of the tests took place in the most innocent of places, and most of the participants are still completely unaware that they have taken part in a scientific study.

Imagine that you walk up to a bank cashpoint. You are just about to place your card in the machine, and out pops a £10 note. Would you take the money and run, or return it to the bank?

The first experiment was designed to discover how a random sample of people would react to this situation. The programme-makers obtained special permission to take over the cashpoint of a well-known British high street bank for the day. They had an engineer remove all the normal machinery, and replace it with a device that dispensed a £10 note whenever people stood in front of the machine. Our first unsuspecting customer walked up to the machine, and, on cue, out popped a £10 note. Secret cameras recorded her every move. The woman proved to be remarkably honest. She immediately took the note into the bank, handed it to the bemused cashier, and explained what had happened. But such honesty proved to be the exception, rather than the rule. Just over two-thirds of people kept the cash, with some returning several times to make the most of the opportunity. Our most dishonest participant returned on twenty occasions.

Why had so many people been prepared to take money that didn't belong to them? Perhaps the use of a cashpoint had skewed the data. People may consider it one thing to retain money given to them by a machine, but quite another actually to take it from another person. Then there was the fact that people thought they were taking money that belonged to a bank. Perhaps our unsuspecting participants saw the money emerging from the cashpoint as an opportunity to get even with an organization that is seen as making mistakes only in its own favour.

To test these ideas, the programme-makers staged a second experiment. This time, people were given money by a person rather than a machine, and the faceless bank was replaced by a friendly shop.

Imagine that you buy a magazine in a shop, and pay with a £5 note. To your surprise, the cashier gives you change for £10. Would you be honest enough to return the money?

To discover the level of public honesty in this scenario, the programme-makers took over a branch of a national newsagents in the north of England, and transformed it into a laboratory for the day. In the first part of the study, the cashier was instructed to give customers too much change. When anyone paid with a £5 note, the cashier gave them change for £10, and whenever someone paid with a £10 note, they received change for £20. As our first unsuspecting customers entered the shop, the team waited to see whether they would be honest enough to own up to their unexpected windfall. Everyone took the money; many left the shop with a sly smile on their face.

As with all research, it was important to rule out other ways of interpreting the results. Perhaps it wasn't that people had been dishonest, but rather (despite smiling) hadn't noticed that they had been given too much change. The study was repeated, but this time the cashier was asked to count the change aloud. The next set of customers entered the shop, and the cashier carefully counted too much change into their hands. Those paying with a £5 note were given change for £10, and those handing over a £10 note received change for £20. All the customers took the money without saying a word.

To further emphasize the mistake, in a penultimate part of the experiment, the cashier was asked to count the excess change into the customer's hand, look slightly confused, and then ask the customer the value of the note that they had used. Surely this time people would be honest enough to own up? Almost no one told the truth. Interestingly, the shoppers often didn't lie straight away, but rather checked that the cashier had no way of actually knowing whether they had used a £10 note or £20 note ('can't you look in the drawer?'), before calling the situation in their favour. Only one person pointed out the cashier's error. In an interview afterwards, he said that he was a Christian, and that Jesus wouldn't have been pleased if he had held on to the cash. Try as they might, none of the team could come up with a way of testing his theory.

In the final part of the study, one of the team stood outside the shop and posed as a market researcher. When someone who had just accepted too much change emerged from the shop, they were asked a few questions about honesty. Did they think journalists were honest? Could the Queen be trusted? Finally, the most important question of all – 'If you were given too much change in a shop, would you own up and return the money?' Until the final question, everyone's answers were fast and clear. No, they didn't trust journalists. Yes, they thought the Queen was honest. Then people suddenly became evasive. Even though they had just committed the dishonest act in question, they produced much longer, and more vague, answers: 'I can't remember the last time that happened to me', 'I don't usually look at my change', 'I never really check my change'. People couldn't bring themselves to be honest even in an anonymous survey.

The findings presented a fascinating, but depressing, view of human nature. Unethical behaviour was alive and well in modern-day Britain. Although the vast majority of people claim to be upstanding citizens, most of us are more than capable of dishonesty if the situation is right.

However, it was not all doom and gloom. A third and final set of studies revealed that when it comes to being selfish or selfless, small and subtle changes make a big difference. The first stage of the newsagent experiment was repeated, but rather than using a branch of a large chain of newsagents, the location was a small corner shop. Like the cashier in the newsagents, the owner of the shop was asked systematically to give his customers too much change – if they paid with a £5 note, they were given change for £10, and if they paid with a £10 note, they were given change for £20. This time, the results were very different. Whereas before, everyone had taken the money and said nothing, now half of the people instantly returned the excess change. It seemed as if it were acceptable to take money from a large company, but not from a small local shop. When interviewed afterwards, many of the honest customers said that it simply wasn't right to steal from someone who was like them. Their comments provide support for one of the key theories that influences when we give, and when we take. It is all to do with the psychology of similarity.

Nixon, horn-honking, and the mad monk of Russia

Richard Nixon made several inadvertent contributions to psychology. In 1960, he took part in the first-ever televised presidential debate. Radio listeners thought that Nixon had won, whilst television viewers gave the verdict to Kennedy. Why? Because Nixon had refused make-up beforehand, and his face appeared sweaty and anxious throughout the debate. Researchers discovered that television viewers focused on what they saw, rather than what they heard, and so came to the opposite decision to radio listeners.[19] Then there is the famed 'Nixon effect'.[20] When giving his resignation speech after the Watergate scandal, Nixon appeared calm and collected. However, researchers analysing his facial expressions noticed a 'furious rate' of eye-blinking (well above fifty blinks per minute), suggesting extremely high levels of anxiety. Subsequent analyses of blink rates during eight televised presidential debates showed that the candidate who blinked the most frequently lost the forthcoming election in seven instances.[21]

Nixon's time in the White House also proved extremely

helpful to researchers examining the psychology of altruism. His stance on Vietnam resulted in some of the largest peace demonstrations of the era. In April 1971, over 200,000 protesters descended on Washington to stage a massive rally. Whilst the media focused on the possible impact of the event for international policy, psychologist Peter Suedfield, and his colleagues from Rutgers University in New Jersey, saw it as an opportunity to carry out a secret study investigating similarity and helping.[22]

A few months before, the researchers had told an actor to grow his hair long and develop a moustache. Towards the start of the rally, they gave him a 'Dump Nixon' placard, and ensured that he was, in the words of their subsequent paper, 'attired in hip garb'. A female experimenter led the actor into the crowd. At a predetermined moment, the actor suddenly sat on the ground, and held his head in his hands pretending to be unwell. The experimenter then approached an unsuspecting genuine protester and started to work her way through a well-rehearsed script.

Initially, she asked the protester if they would assist her unwell friend. If the protester was willing to lend a helping hand, she then asked if they would mind moving her friend away from the crowd. Those that agreed were then asked to help take the friend to the nearest first aid station. If the protester went along with that request, the actor would ask the protester to help him make the seven-mile trip home. Finally, those protesters who indicated that they were prepared to make the trip were asked to provide the necessary bus fare. At this point, the allegedly ill actor made a sudden and unexpected recovery, and the protester was thanked, and told that their assistance was no longer necessary.

To discover the relationship between the help given and similarity, the experimenters repeated the study under a different set of circumstances. This time, the actor was given a 'Support Nixon' sign, had their hair cut, didn't sport the moustache, and lost the 'hip garb' in favour of far more conservative clothing (a sports shirt, slacks, and loafers). The only thing that didn't change was the script – the experimenter and actor went through exactly the same requests as before.

The two conditions yielded very different levels of help. When the actor looked like a peace protester, the peace protesters appeared to be Good Samaritans. Lots of them offered to help, with many going so far as to offer to finance the actor's bus journey home. Some protesters even offered to drive the actor themselves, or, in cases where the protester had no money or car, accompany the actor on a seven-mile walk to his home. The situation was very different when the actor was clean-shaven and not attired in hip garb. Suddenly, the kindly peace protesters were far less willing to help. The need for assistance was exactly the same, but now the actor was one of 'the opposition'.

The study illustrates a very simple, but powerful, concept. We help people who are like us. Decades of experiments involving tomato-ketchup-covered students lying on the street asking for help have shown the same effect time and again. People are at their most altruistic when those in need match their age, background, and fashion sense. It all makes sense from an evolutionary viewpoint. People who look and behave like us are more likely than others to be genetically related to us, or at least from the same tribe, and so are seen as being more deserving of our goodwill.

My favourite experiment into this similarity effect was conducted by Professor Joseph Forgas from Oxford University, and examined the way different European drivers honked their car horns.[23] Forgas's idea combined the three elements that underlie many quirky ideas: it was brilliant, simple, and slightly strange. He had a man and woman drive around Germany, France, Spain, and Italy in a grey VW Beetle. They drove through various towns of roughly the same size, and tried their best to be at the front of traffic when the traffic lights showed red. As the lights turned green, much to the annoyance of the drivers behind them, they simply sat in the car. Actually, that is not quite true. They carefully noted down how the driver directly behind them used his or her horn, including the time elapsed before the first honk, and the duration of honking. This was dangerous work. In a similar experiment conducted a few years before, several motorists had taken out their frustration by ramming the experimenters' vehicles.[24] However, Forgas and his colleagues lived to tell the tale and, perhaps more importantly, analyse their data.

The Italians proved the most impatient, honking their horns, on average, after about 5 seconds. Next were the Spanish at about the 6 second mark. The French came in at about 7 seconds, and the Germans proved the most patient, at about 7.5 seconds.

In this initial part of the study, the experimenter had been keen to ensure that the motorists were not influenced by the nationality of the stationary drivers. For this reason, the Beetle carried a highly salient Australian insignia. According to the researchers, this 'more or less satisfied the requirements of a generic "foreign" car, representing a nation with a presumably neutral national stereotype'. In a second stage of the study, the team sneakily swapped the Australian insignia for a German one, and repeated the procedure. This time, the Italians, Spanish and French all honked their horns much quicker, with the Italians holding off for just 3 seconds, and the Spanish and French venting their anger at about the 4 second mark. In Germany, however, the situation was quite different. Here, the time before horn-honking was extended to almost 8 seconds. Something as simple as an insignia had instigated feelings of similarity, or dissimilarity, and had a significant impact on the time before drivers started to hit their horns.

This is not the only work to employ bumper stickers to illustrate the important role that similarity plays in our lives.

The summer of 1969 saw a series of bloody encounters between the American police and the African American civil rights organization, the Black Panther Party. Frances Heussenstamm was teaching a course in psychology at the California State College during this period, and many of her black students mentioned that they were receiving a large number of traffic citations. Heussenstamm noticed that all of these students had bumper stickers supporting the Panther Party on their cars, and wondered whether the citations were the result of police prejudice or poor driving.[25]

To find out, Heussenstamm asked forty-five other students with exemplary driving records to participate in a rather unusual experiment. The students were asked to drive around with a Panther Party bumper sticker on their vehicles. All the participants signed a statement saying that they would do nothing to attract the attention of the police, and their cars were carefully inspected and found to be in roadworthy condition. In addition, each morning the students made a pledge to drive safely. The first participant received a ticket for 'incorrect lane change' within two hours of starting the experiment. The following day, five more participants received citations for minor offences, such as 'following too closely' and 'driving too slowly'. The participants paid their fines in person after receiving the citation, with one participant receiving his second citation on the way to pay his fine for the first one. Within three weeks, the group received a total of thirty-three citations, whereupon the experiment had to be terminated because Heussenstamm ran out of money to pay for the fines. Heussenstamm reports that when she announced the end of the study 'the remaining drivers expressed relief and went straight to their cars to remove the stickers'. Although the design of the study is far from perfect (Heussenstamm suggests that future work should involve a second group of participants driving around with the bumper sticker 'America – Love It Or Leave It'), the results illustrate how something as simple as a bumper sticker has a large impact on whether people help or hinder others, even when it is their job to be fair and impartial.

Professor Jerry Burger and his colleagues at Santa Clara University in California wondered whether people would take the similarity principle way too far. Could they, for example, be persuaded to help a stranger because the two of them shared a completely meaningless symbol of similarity – the same date of birth?

Burger and his team had volunteers visit his laboratory on the pretence of taking part in an experiment on astrology.[26] The experimenter introduced the volunteer to a second participant (actually an actor working with the experimenter), and handed each a form. The front page of the form asked for various personal details, including their name and date of birth. On 50 per cent of occasions, when the genuine participant completed his or her date of birth, the actor surreptitiously noted it, and filled in the same date on his own form. For the other 50 per cent, the actor deliberately wrote a different date.

The experimenter then asked each 'volunteer' to say their date of birth out loud to ensure that they were given the correct horoscope to assess. Half of the participants discovered an amazing coincidence – they shared the same birthday with the other person! (The other half of the participants, of course, found out that the two of them were born on different days.) The participant and actor rated the accuracy of their respective horoscopes, and then left the laboratory. The volunteer thought that the experiment was over. In fact, it was just about to begin.

As the two walked along the hallway, the actor pulled a four-page essay from his bag, and asked the volunteer if they would mind carefully reading it, and then writing a critique about whether the arguments advanced were convincing. Would those volunteers who believed that they shared a birthday with the actor be more accommodating? About a third of people who thought that they did not share a birthday with the actor agreed to help. In the 'wow, we have the same birthday, what a coincidence' group, almost two-thirds agreed. The simple belief in a shared birthday was enough to persuade people to donate a considerable chunk of their valuable time to a complete stranger.

Professors Finch and Cialdini from Arizona State University have even shown that the same effect causes people to turn a blind eye to other people's crimes and misde-meanours.[27] In their study, participants read a biographical sketch describing the terrible crimes committed by Rasputin, the 'Mad Monk of Russia', and then rated the degree to which they thought Rasputin sounded like a nice chap. Unbeknown to the participants, the experimenters had found out their date of birth beforehand and manipulated the text seen by half of the volunteers to ensure that Rasputin's date of birth matched their own. When participants thought they shared a birthday with the mad monk, they were prepared to overlook his wrongdoings and evil deeds, and found him significantly more likeable.

Tom Desmond, charity boxes and *Medical Center*

Chapter 3 described how Stanley Milgram's innovative exper-iments into the 'small world' phenomenon helped explain why people frequently encounter friends of friends. When not conducting these giant games of pass-the-parcel, Milgram also carried out a considerable amount of research into the psychology of pro-social, and antisocial, behaviour.[28] In the late 1960s, he turned his attention to one of the hottest ques-tions: When it comes to hurting or helping others, to what extent is our behaviour influenced by television? In short, do the programmes we watch create the society in which we live?

The results of several surveys into the amount of vio-lence on television had emphasized the need for this work. In 1971, one researcher found that violent incidents were shown on primetime networks at the rate of eight times an hour. Another survey, conducted a few years later, found that children's programming was 'saturated with violence', with 71 per cent of shows containing at least one violent act.[29] Times haven't changed. One recent survey estimated that by the time children leave primary school, they will have witnessed an average of 8,000 murders, and over 100,000 acts of other violence on television.[30]

Previous work into the topic had tended to involve small-scale, laboratory-based studies in which experimenters showed children violent cartoons, and then carefully counted the number of times they punched a large inflatable figure that just happened to be behind them. Milgram was determined to carry out a highly realistic piece of mass-participation research examining the possible impact of television on the entire American nation.

Funded by a large grant from the broadcaster CBS, Milgram persuaded television scriptwriters to pen different endings to an episode of a hugely popular primetime drama series called *Medical Center* (apparently *Mission Impossible* was considered but rejected because, according to Milgram, it 'regularly depicts such a degree of violence that our experimental act would appear trivial by comparison'). During the episode, a hospital orderly named Tom Desmond lost his job, and so had no way of caring for his sick wife and child. In one of the alternative endings Desmond smashed open several fund-raising collection boxes, stole the money they contained, and was *not* caught by the police. In another version he stole the money, but *was* captured. As a control, the experimenters used a 'neutral' episode of the series that was, according to Milgram, 'romantic, sentimental, and entirely devoid of any violence or antisocial behaviour'. Vincent Sherman, a well-known movie director who had made several highly successful films with Hollywood stars Bette Davis and Errol Flynn, was employed to help make the various versions of the episode.

CBS broadcast the three episodes at different times throughout April 1971. Milgram had devised an elaborate, and ingenious, way of assessing the impact of the different programmes on people's behaviour. Prior to the broadcasts, he mailed letters to thousands of people in New York City and St Louis, stating that they had been selected to take part in a market survey, and asking them to a watch an episode of *Medical Center* at a designated time. They were then invited to complete a simple questionnaire about the characters in the episode, and the commercials shown during the show's break. Respondents were told that after the broadcast they could claim a new radio in return for their participation, and were directed to pick it up from a downtown 'Gift Distribution Center'.

The 'Gift Distribution Centers' were actually fake warehouses staffed with actors, and fitted with hidden cameras. When viewers arrived they walked into an empty office, and encountered a notice saying:

We have no more transistor radios to distribute.
This distribution center is closed until further notice.

The apparent lack of radios, combined with the brusque wording of the notice, was designed to elicit a sense of frustration in participants. The same room also contained a charity gift box on one of the walls. The box was overflowing with cash and would have proved a temptation to anyone of dishonest intent. The experimenters had even carefully placed a one-dollar note dangling from the box to tempt those unwilling actually to break it open. This clever set-up allowed Milgram to discover whether those who had seen Desmond steal from the charity box during the television programme were more likely to engage in criminal behaviour. After a few moments, the participants tried to retrace their steps and leave the building. It was then they discovered that the door that they had used to enter the office was locked, and thus had to follow a series of exit signs. These signs led the participants into a small room, wherein they were met by a clerk who explained that there *were* radios available, and gave the participant their gift.

Almost a thousand people came to the Centers. In New York City, CBS broadcast the neutral episode of *Medical Center*, and the episode in which Tom stole from the charity box and was caught by the police. Nine per cent of the people who had seen the *neutral* programme took the dangling dollar, or broke open the charity box. Did watching Tom steal money and be punished for his crimes increase the likelihood of a theft? In fact, it seemed to make people slightly more honest, as only about 4 per cent of people took the dollar or broke into the box. In St Louis, CBS broadcast the neutral episode and the episode in which Tom's crimes went unpunished. Only about 2 per cent of people who saw the neutral programme behaved in a dishonest way, compared to 3 per cent of those who had seen Tom steal the money and get away with his crime.

Worried that any potential effect might have been diluted by the relatively long delay between people watching the programmes and arriving at the Gift Distribution Centers, Milgram repeated the experiment eliminating the delay. In this new study, people in New York's Times Square area were offered a 'Free Ticket For A Color Television Preview'. Those accepting the offer were taken to a room in a nearby building containing just a television set, a chair, and the charity box. Participants were left alone to watch one of the specially filmed episodes of *Medical Center*, and then secretly observed to see whether they removed money from the charity box. The experiment was not a success. Most of the people accepting the free tickets turned out to be alcoholics, drug addicts, or homeless (with several asking if they were allowed to sleep in the laboratory), and the subsequent levels of antisocial behaviour, including participants urinating on floors and threatening staff, forced the early closure of the experiment. As far as I am aware, this is the only instance of an experiment examining the psychology of antisocial behaviour being terminated owing to antisocial behaviour.

Milgram's highly elaborate, expensive, and extensive studies revealed that the TV programmes had little, if any, impact on public behaviour. The findings caused some controversy, with some arguing that they represented conclusive evidence against any legislation to control television programming, whilst others criticized his methods and promoted the case for censorship.

This mass-participation television experiment was not Milgram's only sojourn into the world of antisocial, and pro-social, behaviour. His other contribution had a much larger impact, and involved devising a method that is still used by psychologists around the globe. The idea was simplicity itself, and concerned the innocent act of inadvertently dropping an envelope in the street.[31]

Envelope-dropping, and the Friends of the Nazi Party

In 1963, Milgram and his research assistants secretly wandered around ten districts in New Haven, Connecticut, dropping 300 envelopes in phone boxes, on pavements, and inside shops. The first line of the address on the envelopes read either 'Friends of the Nazi Party', 'Friends of the Communist Party', or 'Medical Research Associates'. The remaining address lines on all of the envelopes were identical – a post office box in Connecticut. Milgram figured that people would be far more likely to pick up the envelopes and put them in a postbox if they felt some level of support for the organization listed in the address. Milgram was right. About 70 per cent of the envelopes addressed to the 'Medical Research Associates' were returned, compared to just 25 per cent of those addressed to either 'Friends of the Nazi Party' or 'Friends of the Communist Party'.

The findings demonstrated that this simple technique could be used to gauge public opinion without ever having to ask people a single question. In doing so, it was a clever way to find out what they actually thought about an issue, rather than relying on notoriously unreliable surveys and opinion polls.

The technique was not, however, without its problems. Milgram was concerned that scattering so many envelopes addressed to organizations associated with Communists and Nazis might arouse suspicion among both the public and police. In an attempt to avoid such unwanted and unnecessary attention, he contacted the FBI prior to the study, and informed them about the research. It was to little avail. When

Milgram telephoned back after the experiment, the agent said that he couldn't remember Milgram's original call, and hinted that a significant number of agents were now involved in the case. The feds were not the only problem. Milgram also reported that researchers often complained of aching feet after walking the considerable distances needed to ensure a satisfactory distribution of envelopes. The situation was made worse by 'helpful' passers-by frequently spotting an envelope immediately after it had been dropped, picking it up and handing it back to an experimenter.

The technique was, however, sufficiently promising for Milgram to devise and test various ways of overcoming these problems. In one instance, he tried distributing the envelopes from a moving car. To avoid arousing suspicion, this had to happen at night, resulting in envelopes frequently ending up face down in unintended and inappropriate places. Undeterred, on another occasion Milgram hired a light aircraft and dropped hundreds of envelopes over Worcester, Massachusetts. Again, the method proved unsuccessful. Many of the envelopes ended up stuck in trees and lying on rooftops. Worse still, others were swept into the aircraft's aileron, endangering the safety of both the pilot and researcher.

Despite these setbacks, Milgram was to employ the envelope-dropping procedure in several additional studies. In one instance, he measured levels of racial prejudice in predominately white and black neighbourhoods in North Carolina. In another the technique was successfully used to predict the outcome of the 1964 presidential election between Goldwater and Johnson (albeit dramatically underestimating the margin of Johnson's subsequent victory). Milgram also attempted to export the technique to the Far East to investigate the percentage of pro-Mao and pro-Nationalist people living in Hong Kong, Singapore, and Bangkok. Unfortunately, this ambitious project was besieged by unforeseen problems. The experimenter Milgram sent to conduct the study in Singapore was placed on a return plane immediately after arriving in the country because of widespread rioting, and the researcher hired in Hong Kong promptly absconded with Milgram's research funds.

The technique is still used by social psychologists today, and has been employed to examine public opinion about a diverse range of issues, including Clinton's impeachment,[32] gay and lesbian issues, abortion, Arab–Israeli relationships,[33] and the attitudes of Catholics and Protestants towards one another in Northern Ireland. In 1999, school student Lucas Hanft carried out one of the largest studies, dropping 1,600 letters addressed to various fictitious organizations that were pro and anti gay marriage in Manhattan and Nassau County. The results revealed that those living in the city were more liberal than suburbanites. Hanft also experienced many of the same types of problem encountered by Milgram, including, for example, being threatened with arrest for littering.

Over the years, psychologists have employed a modified version of the envelope-dropping technique to measure levels of altruism among different communities and countries. The results of these subsequent studies have helped identify who helps, and when. Some of the most intriguing experiments have investigated a group of people who are often perceived as highly helpful – the deeply religious.

The parable of the Good Samaritan, and other religious myths

Results from work examining religion and altruism suggest that, in general, religious people often give to those in need.[34] Some of the more quirky research in the area, however, has questioned whether such altruism is always the case.

In the 1970s, psychologist Gordon Forbes and colleagues from Millikin University in Illinois wanted to discover which religious groups were the most, and least, helpful.[35] There seemed to be little point in simply asking different sets of churchgoers whether they were good people, as everyone was likely to say yes. Instead, the researchers asked a knowledge-able theologian to identify the ten most liberal and ten most conservative churches in the region. During Sunday services at these churches, the experimenters tiptoed around the outside of the building, dropping letters in doorways and parking lots. They then repeated the procedure at local Catholic churches during mass.

The letters were all sealed and unstamped, and addressed to local residents 'Mr and Mrs Fred Guthrie'. Rather sneak-ily, the experimenters had ensured that letters dropped in the liberal, conservative, and Catholic churches respectively could be identified on the basis of Fred's alleged middle ini-tial. Roughly 40 per cent of letters were returned from each of the three types of church. None of the letters had a stamp on them, and thus people picking up the letters faced a choice. They could either place a stamp on the letter and drop it in a postbox, or send it postage due. The Catholics and liberals came out of the study looking most generous, placing stamps on 89 per cent and 87 per cent of the envelopes respectively. Only 42 per cent of those at the conservative churches were prepared to indulge in this act of kindness, however, with the remaining being returned postage due. As noted by the authors:

> These findings suggest that members of conservative churches are as willing to help strangers as are members of liberal or Catholic churches; yet they are far less willing to spend a few cents to do so.

This study is not the only one to question the altruistic intentions of those claiming to be highly religious. In 1973, Princeton psychologists John Darley and C. Daniel Batson reported a remarkable study into religion and helping.[36] At the start of their experiment, a group of trainee ministers at one of the world's leading institutions for theological education were asked to prepare a sermon based around the parable of the Good Samaritan. According to this well-known biblical story, a man is beaten up by thieves, and left lying in the street. Various priests come across the man, but walk by. Eventually, a Good Samaritan goes out of his way to provide assistance, and the parable ends by urging others to help those in need. After they had made their preparations, the trainee ministers were told that their sermons would be filmed in another building, and were given directions to the new location, and sent on their way. Although they didn't realize it, every step of their journey was being secretly observed by the experimenters.

On the short trip between the two buildings, each participant came across a man (actually an actor) who was clearly in need of some assistance. He was slumped against a doorway, with his head down and eyes closed. As each participant walked past, the actor gave a well-rehearsed groan and two coughs. The experimenters wanted to know whether the trainee ministers would practise what they preached, and help the man. Even though they were on their way to deliver a sermon about the importance of being a Good Samaritan, more than half of the participants walked straight past the man. Some of them actually stepped *over* him. In a slightly modified version of the study, the experimenters told another group of trainee ministers that they needed to get to the second building as soon as possible. Under these circumstances, the level of helping dropped to just 10 per cent. The experiment reveals a great deal about human nature, including the dramatic difference between people's words and actions, and how a fast pace of life can help create an uncaring culture.

Earlier in this chapter I described a series of studies, conducted by the *World In Action* TV programme, examining the honesty of the nation. The programme-makers also carried out a study comparing the honesty of two groups of the most, and least, trusted people in society: priests and second-hand car dealers. According to the results of a recent Gallup Poll, 59 per cent of people rate the clergy as honest, versus only 5 per cent for car salesmen. But do these beliefs really reflect actual honesty? To find out, the team set up a fake soft-furnishings company called 'Honesty', and sent a group of priests and second-hand car dealers a letter from our newly formed organization. The letter thanked them for their recent purchase, and contained a cheque for just over £10 as a refund. All of the recipients would have known that they didn't buy anything from the company, but how many of them would be dishonest enough to cash the cheque? There was very little difference between the two groups, with both the priests and car dealers cashing about 50 per cent of the cheques.

City living

A slight variant of Milgram's envelope-dropping technique has also been used by Robert Levine, from the California State University, to assess kindness across the world.

Levine's initial work investigated whether people helped or not in thirty-six major cities across America.[37] Rather than dropping envelopes in the street, Levine and his team placed stamped addressed envelopes on the windshields of randomly selected cars parked in shopping centres, along with a neat handwritten note saying 'I found this next to your car.' They wanted to see how many of the letters were returned from each area. This test of helping was supplemented by several others. They dropped pens as they walked in front of randomly selected people, and counted how many people picked up the pens and returned them. A perfectly healthy experimenter donned a large leg brace and stood on the street struggling to pick up a pile of magazines that he had dropped whilst a concealed experimenter observed the response of the public. The same experimenter also put on some dark glasses, held a white cane, and took note of how many passers-by helped him cross a busy street.

Levine invested a great deal of time and effort in making these tests as scientific as possible. In the pen-dropping task, for instance, researchers ensured that they were able to walk consistently at a standard speed (1.5 paces per second) towards someone moving in the opposite direction, and rehearsed naturally reaching into their pocket and, without appearing to notice, dropping a pen. When pretending to be blind, the researchers located themselves on street corners that had 'crosswalks, traffic signals and moderate, steady pedestrian flow'. Stepping up to the corner when the lights went green, the experimenters secretly timed how long it was before someone helped them across the road.

Overall, they found that people in small-sized towns in the south-east of the country were most helpful, whereas those living in large towns in the north-east were least likely to assist. Top of the 'helping' list came Rochester in New York; with Houston, Texas, a close second; Nashville, Tennessee, third; and Memphis, Tennessee, fourth. America's least helpful can be found in Patterson, New Jersey, with the second- and third-lowest positions going to New York and Los Angeles respectively.

The results from the lost-letter test proved especially interesting. In New York, the letters were sometimes returned with angry and abusive comments scrawled on them. As Levine notes when describing the experiment in his book *The Geography of Time*:

> Only from New York did I receive an envelope which had its entire side ripped and left open. On the back of the letter the helper had scribbled, in Spanish: '*Hijo de puta iresposable*' – which, translated, makes a very nasty accusation about my mother. Below that was a straight-forward English-language 'F____You'.

In Rochester it was different. One anonymous Good Samaritan wrote a very pleasant note to accompany the lost letter, adding a postscript that is reminiscent of Milgram's original work into small worlds. The envelope asked Levine: 'Are you related to any Levines in New Jersey or Long Island?'

Flushed by the success of their national study, Levine and his colleagues decided to go global.[38] They travelled around the world, visiting capital cities in twenty-three different countries. They dropped more than 400 pens, donned the leg braces over 500 times, and lost about 800 letters. The lost-letter technique proved a cross-cultural nightmare. In Tel Aviv, packages and letters lying on the ground or placed on car windshields, are often associated with bombs, and so were given a wide berth by almost everyone. In El Salvador, they generated suspicion because they often form part of a well-known scam in which a person will pick up the letter, only to find a man standing at their side. The man claims that the letter is his, that it contained some money, that the money is now missing, and so would they like to hand over their own hard-earned cash? Some other countries had no letterboxes or, as in Albania, no reliable postal system. However, despite the difficulties, the researchers persevered and eventually produced the international helping poll.

It proved to be good news for Latin America, with Rio de Janeiro (Brazil) and San José (Costa Rica) heading the list of highly helpful countries. Lilongwe (Malawi) in Africa came in third. Singapore (Singapore), New York (America) and Kuala Lumpur (Malaysia) filled the bottom three places. The differences were far from trivial. In Rio de Janeiro and Lilongwe, 'blind' experimenters were helped across the street on every occasion, whereas in Singapore and Kuala Lumpur they encountered only a 50 per cent success rate. In San José 95 per cent of people helped the experimenter sporting a leg brace to pick up the dropped magazines. In New York, the figure fell to just 28 per cent.

Looking deeper into his data on helping in American cities, Levine and his colleagues discovered that population density provided one of the best predictors of helping. Why should higher population densities lead to less helping? According to one theory, developed by Milgram, people in high-population cities tend to experience a greater amount of 'sensory overload'.[39] They are constantly being bombarded with information from other people, their mobile telephones, traffic, and advertising. As a result, they do what all systems tend to do when receiving too much information – they prioritize, and spend less time dealing with the various sources competing for their attention. Milgram believed that this resulted in people walking past those in need of help, and diverting the responsibility to assist these individuals onto others. All of this creates a paradox, wherein the greater the number of people occupying a space, the greater the sense of loneliness and isolation.

But Levine wasn't just curious about the relationship between the size of a city and the help that citizens provided. He wondered whether helping or not is also determined by the speed of life in the city.

Measuring the pace of life

Eager to put numbers to these seemingly elusive factors, Levine and his co-workers visited thirty-one countries around the world, taking three indicators of the speed of life.[40]

He measured the average walking speed of randomly

selected pedestrians over a 60-foot stretch of pavement, visited various post offices and secretly timed how long it took them to serve a customer buying a single stamp, and took a note of the accuracy of the clocks in fifteen randomly selected downtown banks.

The work was highly methodical. When measuring walking speed, the investigators ensured that all of the locations were flat, free from obstruction, and not especially crowded. Children, those with obvious physical disabilities, and window-shoppers were excluded from the analyses. When timing the speed of service in post offices, the experimenters handed clerks a note written in the native language, to help minimize any potential cross-cultural confusion. Analyses showed that the three measures were all related to one another, suggesting that they did indeed provide an indicator of a city's speed of life.

Levine combined the different indicators into a single measure of speed. The results revealed that Switzerland has the fastest pace of life in the world (their bank clocks showed a discrepancy of just nineteen seconds), with Ireland second and Germany third. Interestingly, eight of the nine fastest countries were from western Europe (Japan broke the total domination of the pole positions by coming fourth). England came fifth, and had the fourth-highest walking speed of the entire list. The only western European country involved in the study not to make it into the top ten was France (it came eleventh, just behind Hong Kong), a result that Levine attributes to the measurements being taken at a time when the country was experiencing one of its hottest summers on record. The three slowest countries were Brazil, Indonesia, and Mexico. The bottom eight positions were all held by countries from Africa, Asia, the Middle East, and Latin America. Within America, Boston proved to be the speediest (just beating New York to pole position), and Los Angeles the most laid-back. The study also revealed more evidence of New Yorkers' rudeness, as it was one of only two cities where the experimenters were insulted by postal clerks (the other was Budapest).

Levine found some evidence that cities with a slower pace of life are more helpful. As predicted by Milgram's 'sensory overload' theory, the more people rush around, the less time they have to devote to factors that are peripheral to their main goals.

This is not the only downside of living in a society with a faster pace of life. In the late 1980s, Levine and his team visited thirty-six cities across America, and compared pace of life with the city's rate of death from coronary heart disease.[41] The hypothesis was simple. People living in faster-paced cities were more likely to resemble the so-called Type A personality. This cluster of traits place great emphasis on urgency, competitiveness, and a general rushing around trying to achieve a great deal in very little time. Type As tend to talk fast, and finish other people's sentences for them. They are often the first to finish at the dinner table, and glance at their watch more frequently than most. Some researchers believe that this mode of living places a large number of stresses and strains on the body. Levine's work showed that cities living life in the fast lane had higher numbers of smokers, and increased rates of coronary heart disease. Further analyses showed that the speed of walking, and the percentage of people wearing watches in each city, were especially good predictors of the problem. Why is there such an unhealthy relationship between these factors? Perhaps Type As are attracted to fast-paced cities. Perhaps living in such speedy places causes people to become Type As. Perhaps it is a combination of the two. Whatever the explanation, the message is clear. In addition to making people less helpful to others, speed kills.

All together now

Levine's global measures of magazine-, pen-, and letter-dropping suggest that population density, and speed of life, are not the only factors that influence levels of helping. Do you care about others, or are you out for yourself? Some psychologists believe that the way in which people answer this question is, to a large extent, culturally determined. Some communities and countries have adopted a set of values that researchers refer to as 'individualism'. These societies stress the needs and rights of the individual, and place less emphasis on rewarding activities that benefit groups of people. At the other end of the spectrum is the collectivist approach, in which people view themselves as part of a larger group (be it family, organization, or an entire society), and tend to reward behaviour that is for the greater good. Levine's results contain tentative evidence that highly individualistic societies (such as America, Britain, and Switzerland) are less caring than collectives (such as Indonesia, Syria, and China). Other work suggests that the effect starts in our early years. When researchers asked 4-year-old children to make up stories about their dolls, the narratives produced by Indonesian children included more friendly and helpful characters than those created by children from America, Germany, or Sweden.[42]

One of the most dramatic studies demonstrating the impact of living in a caring community was carried out by social psychologist Philip Zimbardo.[43] Like Stanley Milgram (who carried out the work into obedience, small worlds, envelope-dropping, and television violence), Zimbardo has also conducted several experiments that have stood the test of time, and is perhaps best known for his now infamous 'prison' study. During this experiment college students randomly allocated to the role of guards in a mock prison behaved in a highly sadistic way towards fellow students allocated the role of prisoners. Such high-profile research is not the only link between Zimbardo and Milgram. As children, the two of them attended the James Monroe High School in the Bronx, New York, and at one stage even sat next to one another in several classes.[44] Also, like Milgram, Zimbardo was interested in the psychology of helping.

His most striking contribution to the area examined the effects of community on antisocial behaviour. Zimbardo secretly filmed what happened when he left a used car unlocked with its bonnet up on a street opposite New York University. After just ten minutes, a passing car stopped and a family got out. The mother quickly removed anything of value from the interior of the car, the father removed the radiator with a hacksaw, whilst their child rooted through the boot. About fifteen minutes later, another two men jacked up the car and removed its tyres. Over the next few hours, other people stripped the vehicle until nothing of value remained, with most of the antisocial behaviour happening in broad daylight. In just two days, Zimbardo secretly filmed more than twenty instances of destruction (mostly committed by middle-class white adults in broad daylight), and the resulting carnage was so bad that two trucks were required to remove the wrecked car from the street.

Zimbardo then left a similar car (again with its bonnet raised) in a location that had a much greater sense of community – opposite Stanford University in Palo Alto, California. In stark contrast to the events that took place in New York City, not one instance of vandalism was recorded over the course of a week. When it started to rain, one passer-by lowered the car's bonnet to protect the motor. When Zimbardo eventually went to remove the car, three people called the police to report that an abandoned car was being stolen.

But how does one create a sense of social responsibility? How do you stop people from thinking about only their own needs and concerns, and move them towards seeing themselves as part of a larger community? The good news is that work conducted by Stanford University psychologists Jonathan Freedman and Scott Fraser suggests that it doesn't take much.[45]

In the first part of their study, a researcher posed as a volunteer worker. They went from door to door in a residential Californian neighbourhood, asking people if they would mind placing a sign in their gardens to help cut speeding in the area. There was just one small problem. It was a very big sign that would completely ruin the look of the person's house and garden. To make the point as vividly as possible, the researcher showed residents a photograph of the large, poorly written sign saying 'DRIVE CAREFULLY' on someone's lawn. It completely dominated the surroundings, concealed much of the front of the house, and completely blocked the doorway. Perhaps not surprisingly, few residents took up the offer.

In stage two of the experiment, the researcher approached a second set of residents and asked them almost exactly the same question. This time, however, the sign was much, much smaller. It was just three inches square and said 'BE A SAFE DRIVER'. It was a small request and almost everyone accepted. Two weeks later, the researcher returned and now asked them to display the bigger sign. This time, 76 per cent of people agreed to place the large ugly placard in their garden.

Why the dramatic change? Freedman and Fraser believe that agreeing to accept the first small sign had a dramatic effect on how residents saw themselves. Suddenly, they were the type of people that helped out. They were good citizens, people who were prepared to make sacrifices for the greater good. So, when it came to making a decision about the big, horrible sign, they were much more likely to say yes. It is a striking example of how to create cooperation. Get people to agree to the small, and it is much easier to persuade them not to worry about the big.

Epilogue

At 1 p.m. on 22 August 2006 I found myself standing outside the General Post Office in Dublin city centre. The building's magnificent facade has six large stone columns. I carefully trundled a surveyor's wheel along the pavement, and discovered that the first and fifth columns were exactly 60 feet apart. Leaning against the fifth column, I pretended to be just another tourist enjoying the summer sun. In reality, I had a stopwatch hidden in my left hand, and was secretly observing people as they walked in front of the building. I was on the lookout for approaching lone pedestrians. Whenever a person walked past the first pillar, I hit the 'start' button on the stopwatch. A few seconds, and exactly 60 feet, later, they would pass me, and I stopped the clock. I then furtively glanced at the stopwatch, and wrote down the time in the battered notebook that has accompanied me on most of my field trips. I was not the only one carrying out this rather strange activity that day. A large team of researchers were making exactly the same measurements in thirty-two countries across the globe.

At the start of this book I described my first quirkology experiment. Conducted in 1985, the study involved approaching people in a railway station with a hidden stopwatch, and having them judge how many seconds had passed since I had introduced myself. Twenty-one years later, I have just carried out my latest piece of research. Like my railway station experiment, this study also involved unsuspecting passers-by and a secret stopwatch. Unlike my earliest work, this was no small-scale affair.

The British Council has offices worldwide, each of which promotes Britain abroad through the arts, education, science, technology, and sporting events. In late 2005, I suggested that we join forces to stage a large-scale cross-cultural experiment examining the pace of life across the globe. Following on from Robert Levine's innovative research into the topic (described in chapter 6), we aimed to measure worldwide walking speeds in the twenty-first century, and, by comparing the results with Levine's data from the early 1990s, to discover whether people are moving faster than ever before.

On the same day that I was in Dublin, our research teams around the globe ventured into city centres armed with a stopwatch, measuring tape, paper, and a pen.[1] Following in the footsteps of Levine, they found a busy street with a wide pavement that was flat, free from obstacles, and sufficiently uncrowded to allow people to walk along at their maximum speed. Between 11.30 a.m. and 2.00 p.m. local time, they recorded how long it took thirty-five men and thirty-five women to walk along a 60-foot stretch of pavement. They only monitored adults who were on their own, and ignored anyone holding a mobile telephone conversation or struggling with shopping bags.

From Paris to Prague, Singapore to Stockholm, the researchers measured the pace of life in major cities across thirty-two countries. The survey covered many of the cities documented by Levine, plus several not included in his 1994 study. The overall rankings for the different cities, and countries, are shown overleaf.[2]

Fastest

Rank	City	Country
1	Singapore	Singapore
2	Copenhagen	Denmark
3	Madrid	Spain
4	Guangzhou	China
5	Dublin	Ireland
6	Curitiba	Brazil
7	Berlin	Germany
8	New York	United States of America
9	Utrecht	Netherlands
10	Vienna	Austria
11	Warsaw	Poland
12	London	United Kingdom
13	Zagreb	Croatia
14	Prague	Czech Republic
15	Wellington	New Zealand
16	Paris	France
17	Stockholm	Sweden
18	Ljubljana	Slovenia
19	Tokyo	Japan
20	Ottawa	Canada
21	Harare	Zimbabwe
22	Sofia	Bulgaria
23	Taipei	Taiwan
24	Cairo	Egypt
25	Sana'a	Yemen
26	Bucharest	Romania
27	Dubai	United Arab Emirates
28	Damascus	Syria
29	Amman	Jordan
30	Bern	Switzerland
31	Manama	Bahrain
32	Blantyre	Malawi

Slowest

On the same day, we also sent out teams to each of the capital cities within the United Kingdom. Londoners were moving the fastest, taking an average of 12.17 seconds to cover 60 feet. Next were people in Belfast, who covered the same distance in 12.98 seconds. Third place went to Edinburgh, with an average of 13.29 seconds, and the slowest walkers were found in Cardiff, who clocked up an average time of 16.81 seconds. These differences may not seem much, but they all add up. Whilst it would take Londoners approximately eleven days of continuous walking to cover the 874 miles from Land's End to John o'Groats, people from Cardiff would take just under fifteen days to make the same journey.

By comparing the sixteen cities that were in Levine's work and our own, we were able to determine whether the pace of life was increasing. I had travelled to Ireland because in Levine's survey Dubliners proved to be the fastest, covering the 60-foot distance in an average of 11.13 seconds. In 2006, the situation was roughly the same, with my results showing an average of 11.03 seconds. This pattern was repeated for a few of the other cities towards the top of Levine's table, including Tokyo, New York, London, and Paris. But what about the slower cities in Levine's list? In 1994, people in Bucharest took an average of 16.72 seconds, whilst in 2006 they completed the distance in an average of 14.36. People in Vienna were now going 2 seconds faster, with their original average time of 14.08 cut to 12.06. The same pattern emerged in Sofia, Prague, Warsaw, and Stockholm. The biggest changes were found in Guangzhou and Singapore. People in both of these cities now take, on average, 4 seconds less to walk 60 feet than in the early 1990s, suggesting that the pace of life there is increasing around four times faster than in many parts of the world. In the early 1990s, the over-all average walking speed in the sixteen countries was 13.76 seconds. In 2006, this figure had fallen to 12.49 seconds. Our global walking experiment suggested that people around the world are indeed moving faster than ever.

This increase has been achieved in just over ten years. Projected forward, the results suggest that by 2021, people will be covering the same distance in almost no time at all. By 2040, they will arrive at their destination several seconds *before* they have set off. In view of Milgram's 'sensory overload' hypothesis (described in chapter 6), the findings suggest that people in these cities are likely to be less caring, more focused on what matters to them, and more isolated from one another than ever before. Given the important role that these factors play in creating caring communities, some might argue that city dwellers around the world are now moving faster than the speed of life. It is a chilling thought, especially as the recent United Nations report, *State of the World's Cities 2006/7*, concluded that, for the first time in history, more people are now living in cities than in the countryside.

After completing my Dublin measurements, I closed my notebook, walked away from the General Post Office (at 5.45 feet per second), and reflected on twenty-one years of examining the quirky side of life. In addition to diverse topics like testing the lie-detection skills of nations and uncovering the psychology of chat-up lines and personal ads, it has produced many wonderful and highly memorable moments, including spending nights in apparently haunted houses, performing unfunny stand-up routines in comedy clubs, having my doctoral students dress as giant chickens, watching a 4-year-child beat the stock market, and feeling the heat emanate from a 60-foot bed of white-hot coals (we didn't measure the speed of walking during that study, but I suspect it would have made Dubliners look slow). But no scientific work exists in a vacuum. My research has built upon studies conducted by academics who have also dared to explore the backwaters of human behaviour. Researchers who have suffered for their science by secreting themselves in supermarkets and bowling alleys, applied voltages to the corpses of murderers, stalled cars at traffic lights, and spent countless hours trawling through millions of death records. Together, quirkologists have given us important insights into many areas of psychology, including deception, superstition, and altruism. Such work has also helped uncover the secret psychology that underlies our everyday lives, and illustrated just how fascinating those everyday lives really are. Or as Arthur Conan Doyle so eloquently put it in his book *Study in Scarlet*: 'Life is infinitely stranger than anything the mind of man could invent.'

For over a hundred years, a small group of highly dedicated researchers have been studying people like you. To date, their work has only scratched the surface of the fascinating phenomenon that is your life. My hope is that this book will help quirkology move from the margins to the mainstream, and that studying the unusual will become surprisingly commonplace. I hope that my fellow academics will be encouraged to carry out more work that is both interesting and unusual: that they will, for instance, discover whether blondes really do have more fun, why we daydream, the relationship between people's personality and their mobile ringtone, why some people are more likeable than others, whether ventriloquists have multiple personalities, whether wearing school uniform makes children less creative, and why we cry when we are happy. In short, I dream of a world packed full of researchers examining the more offbeat and quirky aspects of life. Next time someone stops you in the street and asks for the correct time, or you are stuck behind a car that has just stalled at traffic lights, or you find a £20 note on the ground, beware. There may be far more going on than you suspect.

Towards the Worldwide Eradication of 'FTSE-itis'

Surveys show that 87 per cent of the population suffer from 'FTSE-itis' – the entirely rational fear of being trapped in dull conversations at dinner parties.[1] To help alleviate such suffering, I recently held a series of 'experimental' dinner parties. Before being allowed access to any food, each of my guests had to rate a long list of factoids derived from the studies described in this book, on a scale from 1 ('Whatever') to 5 ('Really? When does it come out in paperback?'). I then used the data to identify the factoids that were most likely to provoke good conversation at even the dullest of gatherings.

Here are the factoids that took tenth place upwards:

10 People asked to write down a few words describing a university professor answer more Trivial Pursuit questions correctly than those describing a football hooligan.

9 Women's personal ads would attract more replies if they were written by a man. The opposite is not true of men's ads.

8 The *Mona Lisa* seems enigmatic because Leonardo da Vinci painted her so that her smile appears more striking when people look at her eyes than at her mouth.

7 Women van drivers are more likely than others to take more than ten items through the express lane in a supermarket, break speed limits, and park in restricted areas. (This one proved especially popular with women.)

6 Some seemingly ghostly experiences, such as feeling an odd sense of presence, are actually due to low-frequency sound waves produced by the wind blowing across an open window. (This received the top score from men.)

5 Words containing the 'K' sound – such as duck, quack, and Krusty the Clown – are especially likely to make people laugh.

4 People born during the summer are luckier than those born in the winter – temperature differences around the time of birth makes summer-borns more optimistic and open to opportunities.

In third place came work relating to the language of lying:

The best way of detecting a lie is to listen rather than look – liars say less, give fewer details, and use the word 'I' less than people telling the truth.

The factoid placed second continued the deception theme, and was all about fake smiles:

The difference between a genuine and a fake smile is all in the eyes – in a genuine smile, the skin around the eyes crinkles; in a fake smile it remains much flatter.

The number one factoid was that curious fact about sweater-wearing and dog faeces:

> People would rather wear a sweater that has been dropped in dog faeces and not washed, than one that has been dry-cleaned but used to belong to a mass murderer.

So, next time you go to a dinner party, arm yourself with one or more of these scientifically proven ways of creating interesting conversation.

Together, we can eradicate 'FTSE-itis'.

The Big Secret Experiment

In the true spirit of quirkology, this book has an unusual two-part study built into it. This study will eventually reveal a great deal about the personality of readers. By buying this book, you have participated in the first part of the study. The second part involves completing a short personality questionnaire at *www.quirkology.com*.

All the information you supply will remain confidential, and we need as many participants as possible for the study to be a success.

It is possible that by the time you read this, the experiment will have been completed. If this is the case, the results will be available at *www.quirkology.com*.

Acknowledgements

This book has its roots in a chance conversation with science writer and *Scientific American* columnist, Michael Shermer. In November 2005, Michael kindly arranged for me to speak at California Institute of Technology. Whilst chatting on the way back to my hotel, the idea of my writing a book about my unusual experiments in psychology emerged. Thank you, Michael. Without that conversation, this book might never have happened.

Various organizations have funded, and helped carry out, many of the studies described here. First and foremost, I wish to thank the University of Hertfordshire for supporting my work over the years. I would like to thank Sue Hordijenko, Jill Nelson, Nick Hillier, Craig Brierley, and Paul Briggs from the British Association for the Advancement of Science for your invaluable work during the financial astrology experiment and LaughLab. My thanks also to Simon Gage, Tracy Foster, and Pauline Mullin from the Edinburgh International Science Festival for helping to stage Born Lucky, and for conducting the studies exploring the science of smiling and speed dating. Thanks also to Katie Smith, and the team from the Cheltenham Festival of Science, for helping to arrange the 'small world' study. Similarly, my thanks to Karen Hartshorn, director of the New Zealand International Science Festival, for helping to conduct Born Lucky 2, and for arranging the smiling exhibition–experiment at the Dunedin Public Art Gallery. My thanks to the British Council for funding my trip to New Zealand, and to Felicity Connell for doing such a wonderful job of looking after me whilst I was there. I would also like to thank Michael White from the British Council for organizing the 'global pace of life' study, and to the teams of researchers who found the time to measure the speed of walking around the world.

Much of the work described here has involved the media, and I have been lucky enough to work with several talented journalists and programme-makers over the years. My thanks to Penny Park and Jay Ingram from the *Daily Planet* for collaborating on so many studies, and for persuading my childhood hero, Leslie Nielsen, to participate in one of our experiments. Also, thanks to John Zaritsky and the team for creating many happy memories when we filmed the *No Kidding* documentary on LaughLab, and Isobel Williams from Bite Yer Legs for the surreal experience of watching people offer to pay several pounds for a worthless brass curtain ring. Special thanks are due to Roger Highfield, Science Editor of the *Daily Telegraph*, and writer Simon Singh. Roger, thank you for introducing me to the heady world of science communication, and for turning so many of my ideas into reality. As you remind me every time we meet, without you I would be nothing. Simon, thank you for making such a great job of the experiment with Sir Robin Day, and for your invaluable advice and expertise over the years. Without you, Roger would be nothing.

Thanks are also due to my colleagues and collaborators. To Matthew Smith, who carried out the lottery experiment, and spent lots of time doing secret stuff in the dark during fake seances. To Emma Greening, for sending out all those parcels in the 'small world' study, ghost-hunting, exploring the psychology of suggestion, vetting thousands of jokes, and still finding the energy to laugh. To Sarah Woods, for having her brain scanned, measuring the pace of life in London, and not taking the LaughLab blonde jokes personally. To Ciarán O'Keeffe, for dressing up as a giant chicken, and for exploring some of Britain's least haunted locations with me. To Adrian Owen, for helping out with brain scanning, and finding the origin of the world's funniest joke. To all of the Infrasonic team (Sarah Angliss, Ciarán O'Keeffe, Richard Lord, Dan Simmon, and GéNIA) for managing to have such a good time whilst carrying out a study that has inspired so many. To Jim Houran and Jayanti Chotai for sharing your invaluable expertise on ghostly experiences, speed dating, and chronopsychology. To Karen, for helping with the speed-dating experiment and for allowing us to use your photograph on the cover of the book. To Peter, for allowing me to reproduce your wonderfully instructive fake and genuine smiles. To Brian Fischbacher for taking such great photographs of Karen and Peter. To Clive Jeffries, for spending so much time in the dark during the seance studies, and for providing such insightful feedback on the book. And to Andy Nyman, for doing such a convincing job of talking to the dead and for making me laugh so much – you deserve any success that may eventually come your way.

This book would not have been possible without the guidance and expertise of my agents Patrick Walsh and Emma Parry, and editors Jason Cooper, Richard Milner, and Joann Miller. Special thanks also to my wonderful colleague and collaborator, Caroline Watt. You have helped design and conduct almost all the studies described here, provided much-needed support when the going got tough, and have given far beyond the call of duty. Thank you.

Finally, my thanks to the researchers who have carried out the hundreds of slightly strange studies described, and the millions of participants who have contributed to this work. Without you, the book would have been completely different, and much shorter.

Notes

Introduction

1 M. Brookes – *Extreme measures: The Dark Visions and Bright Ideas of Francis Galton*. Bloomsbury: London, 2004.

2 F. Galton – 'Statistical inquiries into the efficacy of prayer', *Fortnightly Review* #68, pages 125–35. 1872.

3 id. – *The Art of Travel*, page 209. John Murray: London, 1855.

4 I. Farkas, D. Helbing & T. Vicsek – 'Mexican waves in an excitable medium', *Nature* #419, pages 131–2. 2002.

5 L. Standing – 'Learning 10,000 pictures', *Quarterly Journal of Experimental Psychology* #25, pages 207–22. 1973.

6 R. Sommer – 'The personality of vegetables: Botanical metaphors for human characteristics', *Journal of Personality* #56(4), pages 665–83. 1988.

7 J. Trinkaus – 'Wearing baseball-type caps: An informal look', *Psychological Reports* #74(2), pages 585–6. 1994.

8 R. B. Cialdini & D. A. Schroeder – 'Increasing compliance by legitimizing paltry contributions: When even a penny helps', *Journal of Personality and Social Psychology* #34, pages 599–604. 1976.

9 R. A. Craddick – 'Size of Santa Claus drawings as a function of time before and after Christmas', *Journal of Psychological Studies* #12, pages 121–5. 1961.

1. What does your date of birth *really* say about you?

1 G. Dean – 'Astrology' in G. Stein (ed.), *The Encyclopedia of the Paranormal*, pages 47–99. Prometheus Books: Amherst. 1996.

Notes

2 D. T. Regan – *For the Record: From Wall Street to Washington*. Harcourt Brace: New York, 1988.

— id. – *What Does Joan Say?: My Seven Years as White House Astrologer to Nancy and Ronald Reagan*. Carol Publishing Group: New York, 1990.

3 J. Chapman – 'How a girl of four trounced a top investor and a stargazer at playing the stock market', *Daily Mail*, 21 March 2001, pages 2–3.

4 M. Nichols – 'An investor, an astrologer, and a girl, 4, played the market. Guess who won?', *Scotsman*, 24 March 2001, page 5.

5 T. Teeman – 'Girl shows money game is child's play', *The Times*, 24 March 2001.

6 G. Rollings – 'McNuggets of wisdom from the shares ace aged four', *Sun*, 24 March 2001, page 50.

7 Ibid.

8 T. Utton – 'Girl of five beats the stock market experts (again)', *Daily Mail*, 14 March 2002, page 43.

9 J. Mayo, O. White & H. J. Eysenck – 'An empirical study of the relation between astrological factors and personality', *Journal of Social Psychology* #105, pages 229–36. 1978.

10 Editorial – 'British scientist proves basic astrology theory', *Phenomena*, 1 April 1977, page 1.

11 H. J. Eysenck & D. K. B. Nias – *Astrology: Science or Superstition?*. Pelican: London, 1988.

12 H. B. Gibson – *Hans Eysenck: The Man and His Work*, page 210. Peter Owen: London, 1981.

13 G. Jahoda – 'A note on Ashanti names and their relationship to personality', *British Journal of Psychology* #45, pages 192–5. 1954.

14 M. Gauquelin – *Dreams and Illusions of astrology*. Glover & Blair Ltd: London, 1979.

15 G. Dean & I. W. Kelly – 'Is astrology relevant to consciousness and psi?', *Journal of Consciousness Studies* #10(6–7), pages 175–98. 2003.

16 For an overview of this work, see: G. Dean, A. Mather & I. W. Kelly – 'Astrology' in G. Stein (ed.), *The Encyclopedia of the Paranormal*, pages 47–99. Prometheus Books: Amherst, 1966.

17 V. Muhrer – 'Astrology on Death Row', *The Indian Skeptic* #11, pages 13–19. 1989.

Notes

18 B. R. Forer – 'The fallacy of personal validation: A classroom demonstration of gullibility', *Journal of Abnormal Psychology* #44, pages 118–21. 1949.

19 P. E. Meehl – 'Wanted – A good cookbook', *American Psychologist* #11, pages 263–72. 1956.

20 D. H. Dickson & I. W. Kelly – 'The "Barnum Effect" in personality assessment: A review of the literature', *Psychological Reports* #57, pages 367–82. 1985.

21 M. Gauquelin – *Dreams and Illusions of astrology.* Glover & Blair Ltd: London, 1979.

22 S. J. Blackmore – 'Probability misjudgment and belief in the paranormal: A newspaper survey', *British Journal of Psychology* #88, pages 683–9. 1997.

23 M. Hamilton – 'Who believes in astrology? Effects of favorableness of astrology derived personality descriptions on acceptance of astrology', *Personality and Individual Differences* #31, pages 895–902. 2001.

24 M. Siffre – *Beyond Time.* McGraw-Hill: New York, 1964.

25 S. S. Campbell & P. J. Murphy – 'Extraocular circadian phototransduction in humans', *Science* #279, page 396. 1998. This finding has been challenged in the following paper: K. P. Wright & C. A. Czeisler – 'Absence of circadian phase resetting in response to bright light behind the knees', *Science* #297, page 571. 2002.

26 A. Dudink – 'Birth date and sporting success', *Nature* #368, page 592. 1994.

27 R. H. Barnsley, A. H. Thompson & P. E. Barnsley – 'Hockey success and birthdate: The relative age effect', *Canadian Association for Health, Physical Education, and Recreation Journal* #51, pages 23–8. 1985.

— S. Edwards – letter to the Editor, 'Born too late to win?' *Nature* #370, page 186. 1994.

— J. Musch & R. Hay – 'The relative age effect in soccer: Cross-cultural evidence for a systematic discrimination against children born late in the competition year', *Sociology of Sport Journal* #16, pages 54–64. 1999.

— A. H. Thompson, R. H. Barnsley & G. Stebelsky – 'Born to play ball: The relative age effect and major league baseball', *Sociology of Sport Journal* #8, pages 146–51. 1991.

28 This work is summarized in R. Wiseman – *The Luck Factor*. Random House: London, 2004.

29 J. Chotai & R. Wiseman – 'Born lucky? The relationship between feeling lucky and month of birth', *Personality and Individual Differences* #39, pages 1451–60. 2005.

30 J. Chotai, T. Forsgren, L. G. Nilsson & R. Adolfsson – 'Season of birth variations in the temperament and character inventory of personality in a general population', *Neuropsychobiology* #44, pages 19–26. 2001.

31 S. Dickert-Conlin & A. Chandra – 'Taxes and the timing of births', *Journal of Political Economy* #107(1), pages 161–77. 1999.

32 A. A. Harrison, N. J. Struthers & M. Moore – 'On the conjunction of National Holidays and reported birthdates: One more path to reflected glory?', *Social Psychology Quarterly* #51(4), pages 365–70. 1988.

33 H. J. Eysenck & D. K. B. Nias – *Astrology: Science or Superstition?*. Pelican: London, 1998.

34 For a readable introduction to this controversy, see: G. Dean – 'Is the Mars Effect a social effect? A re-analysis of the Gauquelin data suggests that hitherto baffling planetary effects may be simple social effects in disguise,' *Skeptical Inquirer* #26(3), pages 33–8. 2002.

— S. Ertel – 'The Mars Effect cannot be pinned on cheating parents – Follow-up', *Skeptical Inquirer* #27(1), pages 57–8. 2003.

— G. Dean – 'Response to Ertel', *Skeptical Inquirer* #27(1), pages 57–60, 65. 2003.

35 D. P. Phillips & D. G. Smith – 'Postponement of death until symbolically meaningful occasions', *Journal of the American Medical Association* #263, pages 1947–51. 1990.

36 D. P. Phillips, C. A. Van Voorhees & T. E. Ruth – 'The birthday: Lifeline or deadline?' *Psychosomatic Medicine* #54, pages 532–42. 1992.

37 For a review of this data and debate, see: J. A. Skala & K. E. Freedland – 'Death takes a raincheck', *Psychosomatic Medicine* #66, pages 382–6. 2004.

38 S. A. Everson, D. E. Goldberg, G. A. Kaplan, R. D. Cohen, E. Pukkala, J. Tuomilehto & J. T. Salonen – 'Hopelessness and risk of mortality and incidence of myocardial infarction and cancer', *Psychosomatic Medicine* #58, pages 113–21. 1996.

Notes

39 W. Kopczuk & J. Slemrod – 'Dying to save taxes: Evidence from estate-tax returns on the death elasticity', *Review of Economics and Statistics* #85(2), pages 256–65. 2003.

2. Trust everyone, but always cut the cards

1 M. D. Morris – 'Large scale deceit: Deception by captive elephants?', in R. W. Mitchell & N. S. Thompson (eds), *Deception: Perspectives on human and nonhuman deceit*, pages 183–92. State University of New York Press: New York, 1986.
2 Information about Koko and Michael can be found at: *www.koko.org.*
3 A full transcript of this conversation is available at: *www.koko.org/world/talk_aol.html.* Online Host content: Copyright 1998–2006 AOL LLC. Used with permission.
4 H. L. Miles – 'How can I tell a lie? Apes, language, and the problems of deception' in R. W. Mitchell & N. S. Thompson (eds), *Deception: Perspectives on human and nonhuman deceit*, pages 245–66. State University of New York Press: New York, 1986.
5 M. Lewis – 'The development of deception' in M. Lewis & C. Saarni (eds), *Lying and deception in everyday life*, pages 90–105. The Guilford Press: New York, 1993.
6 P. Ekman – *Telling lies: Clues to deceit in the marketplace, politics, and marriage.* W. W. Norton & Company: New York, 1985.
7 R. Highfield – 'How age affects the way we lie', *Daily Telegraph*, page 26. 25 March 1994.
8 This work is reviewed in A. Vrij – *Detecting lies and deceit.* John Wiley & Sons: Chichester, 2000.
9 R. G. Hass – 'Perspective-taking and self-awareness: Drawing an E on your forehead', *Journal of Personality and Social Psychology* #46, pages 788–98. 1984.
10 R. Wiseman – The MegaLab truth test', *Nature* #373, page 391. 1995.
11 This work is reviewed in: A. Vrij – *Detecting 37*12.16=/36=lies and deceit.* John Wiley & Sons: Chichester, 2000.
12 Cited in B. M. DePaulo & W. L. Morris – 'Discerning lies from truths: Behavioural cues to deception and the indirect pathway of

281

intuition' in P.A Granhag & L. A. Stromwall (eds), *The Detection of Deception in Forensic Contexts*, pages 15–40. Cambridge University Press: Cambridge, 2004.

13 P. Ekman & M. O'Sullivan – 'Who can catch a liar?', *American Psychologist* #46(9), pages 913–20. 1991.

14 The global deception research team – 'A world of lies', *Journal of Cross-Cultural Psychology* #37(1), pages 60–74. 2006.

15 This work is reviewed in: A. Vrij – *Detecting lies and deceit*. John Wiley & Sons: Chichester, 2000.
    And in: B. M. DePaulo & W. L. Morris – 'Discerning lies from truths: Behavioural cues to deception and the indirect pathway of intuition' in P. A. Granhag & L. A. Stromwall (eds), *The Detection of Deception in Forensic Contexts* pages 15–40. Cambridge University Press: Cambridge, 2004.

16 G. Littlepage & T. Pineault – 'Verbal, facial, and paralinguistic cues to the detection of truth and lying', *Personality and Social Psychology* #4(3), pages 461–4. 1978.

17 R. E. Kraut & R. E. Johnston – 'Social and emotional messages of smiling: An ethological approach', *Journal of Personality and Social Psychology* #37, pages 1539–53. 1979.

18 The photographs used in this study were originally taken for a similar online experiment carried out in collaboration with the Edinburgh International Science Festival.

19 C. Landis – 'Studies of emotional reactions: II. General behavior and facial expression', *Journal of Comparative Psychology* #4, pages 447–509. 1924.

20 M. S. Livingstone – 'Is it warm? Is it real? Or just low spatial frequency?' *Science* #290, page 1299. 2000.

21 A. Parent – 'Giovanni Aldini: From animal electricity to human brain stimulation', *The Canadian Journal of Neurological Sciences* #31, pages 576–84. 2004.

22 G. T. Crook – *The Complete Newgate Calendar, Volume Four*. The Navarre Society: London, 1926.

23 'Horrible phenomena! – Galvanism', *Scotsman*, 11 February 1819.

24 G. B. Duchenne de Boulogne – *The Mechanism of Human Facial Expression*. Cambridge University Press: New York, 1990. Reprinting of original 1862 edition.

25 P. Ekman & W. V. Friesen – *The Facial Action Coding System*. Consulting Psychologists' Press: Palo Alto, 1978.

26 D. D. Danner, D. A. Snowdon & W. V. Friesen – 'Positive emotions in early life and longevity: Findings from the nun study', *Journal of Personality and Social Psychology* #80, pages 804–13. 2001.

27 L. A. Harker & D. Keltner – 'Expressions of positive emotion in women's college yearbook pictures and their relationship to personality and life outcomes across adulthood', *Journal of Personality and Social Psychology* #80, pages 112–24. 2001.

28 E. F. Loftus – *Eyewitness Testimony*. Harvard University Press: Cambridge, 1979.

29 K. A. Wade, M. Garry, J. D. Read & D. S. Lindsay – 'A picture is worth a thousand lies: Using false photographs to create false childhood memories', *Psychonomic Bulletin and Review* #9, pages 597–603. 2002.

30 K. A. Braun, R. Ellis & E. F. Loftus – 'Make my memory: How advertising can change our memories of the past', *Psychology and Marketing* #19, pages 1–23. 2002.

31 E. F. Loftus & J. E. Pickrell – 'The formation of false memories', *Psychiatric Annals* #25, pages 720–5. 1995.

32 I. E. Hyman, T. H. Husband & F. J. Billings – 'False memories of childhood experiences', *Applied Cognitive Psychology* #9, pages 181–95. 1995.

33 J. Jastrow – 'Psychological notes upon sleight-of-hand experts', *Science*, pages 685–9. 8 May 1896.

34 R. Wiseman & E. Greening – 'It's still bending': Verbal suggestion and alleged psychokinetic metal bending. *British Journal of Psychology* #96(1), pages 115–27. 2005.

35 R. Wiseman, E. Greening & M. Smith – 'Belief in the paranormal and suggestion in the seance room', *British Journal of Psychology* #94(3), pages 285–97. 2003.

3. Believing six impossible things before breakfast

1 N. Lachenmeyer – *13: The World's Most Popular Superstition*. Profile Books: London, 2004.

2 J. McCallum – 'Green cars, black cats, and lady luck', *Sports Illustrated* #68, pages 86–94. 8 February 1988.

3 D. W. Moore – 'One in four Americans superstitious', Gallup Poll News Service, 13 October 2000.

4 S. Epstein – 'Cognitive-experiential self theory: Implications for developmental psychology' in M. Gunnar & L. A. Sroufe (eds), *Self-processes and development. Minnesota symposia on child psychology*, vol. 23, pages 79–123. Erlbaum: Hillsdale, 1993.

5 T. Radford – 'If you aren't born lucky, no amount of rabbits' feet will make a jot of difference', *Guardian*, page 15. 18 March 2003.

6 S. E. Peckham & P. G. Bhagwat – *Number 13: Unlucky/Lucky for Some*. Peckwat Publications: New Milton, Hampshire, 1993.

7 D. P. Phillips, G. C. Liu, K. Kwok, J. R. Jarvinen, W. Zhang & I. S. Abramson – 'The *Hound of the Baskervilles* effect: Natural experiment on the influence of psychological stress on timing of death', *British Medical Journal* #323, pages 1443–6. 2001.

8 G. Smith – 'Scared to death?', *British Medical Journal* #325, pages 1442–3. 2002.

— N. S. Panesar, N. C. Y. Chan, S. N. Li, J. K. Y. Lo, V. W. Y. Wong, I. B. Yang & E. K. Y. Yip – 'Is four a deadly number for the Chinese?', *Medical Journal of Australia* #179(11/12), pages 656–8. 2003.

9 T. J. Scanlon, R. N. Luben, F. L. Scanlon & N. Singleton – 'Is Friday the 13th bad for your health?', *British Medical Journal* #307, pages 1584–6. 1993.

10 S. Näyhä – 'Traffic deaths and superstition on Friday the 13th', *American Journal of Psychiatry* #159, pages 2110–11. 2002.
      This work has generated the following debate: I. Radun & H. Summala – 'Females do not have more injury road accidents on Friday the 13th', *BMC Public Health* #4(1), page 54. 2004.

— D. F. Smith – 'Traffic accidents and Friday the 13th', *American Journal of Psychiatry* #161(11), page 2140. 2004.

— S. Näyhä – 'Dr Näyhä replies', *American Journal of Psychiatry* #161, page 2140. 2004.

11 K. Kaku – 'Increased induced abortion rate in 1966, an aspect of a Japanese folk superstition', *Annals of Human Biology* #2 (2), pages 111–15. 1975.

12 K. Kaku & Y. S. Matsumoto – 'Influence of a folk superstition on

fertility of Japanese in California and Hawaii, 1966', *American Journal of Public Health*, #65(2), pages 170–4. 1966.

13 K. Kaku – 'Were girl babies sacrificed to a folk superstition in 1966 in Japan?', *Annals of Human Biology* #2(4), pages 391–3. 1975.

14 K. Hira, T. Fukui, A. Endoh, M. Rahman, M. Maekawa – 'Influence of superstition on the date of hospital discharge and medical cost in Japan: Retrospective and descriptive study', *British Medical Journal* #317, pages 1680–3. 1998.

15 D. O'Reilly & M. Stevenson – 'The effect of superstition on the day of discharge from maternity units in Northern Ireland: A Saturday flit is a short sit', *Journal of Obstetrics and Gynecology* #20, pages 139–41. 2000.

— E. M. Keane, P. O'Leary & J. B. Walsh – 'Saturday flit, short sit: A strong influence of a superstition on the timing of hospital discharges?', *Irish Medical Journal* #90, page 28. 1997.

16 P. Haining – *Superstitions*. Sidgwick & Jackson Ltd: London, 1979.

17 M. D. Smith, R. Wiseman & P. Harris – 'Perceived luckiness and the UK National Lottery', *Proceedings of the 40th Annual Convention of the Parapsychological Association*, pages 387–98. UK, 1997.

18 M. Levin – 'Do Black Cats Cause Bad Luck?' Winner of the Joel Serebin Memorial Essay Contest organized by the New York Area Skeptics. Available at *www.petcaretips.net/black_cat_luck*.

19 W. Coates, D. Jehle & E. Cottington – 'Trauma and the full moon: A waning theory', *Annals of Emergency Medicine* #18, pages 763–5. 1989.

— A review of other potential lunacy effects can be found in J. Rotton & I. W. Kelly – 'Much ado about the full moon: A meta-analysis of lunar-lunacy research', *Psychological Bulletin*, #97(2), pages 286–306. 1985.

20 D. F. Danzl – 'Lunacy', *Journal of Emergency Medicine* #5(2), pages 91–5. 1987.

21 A. Ahn, B. K. Nallamothu & S. Saint – '"We're jinxed" – are residents' fears of being jinxed during an on-call day founded?', *American Journal of Medicine* #112(6), page 504. 2002.

22 P. Davis & A. Fox – 'Never say the "Q" word', *StudentBMJ* #10, pages 353–96. 2002.

23 Much of the information in this section is discussed in
N. Lachenmeyer – *13: The World's Most Popular Superstition.*
Profile Books: London, 2004.

24 R. G. Ingersoll – 'Toast: The superstitions of public men', Thirteen
Club Dinner, 13 December 1886.

25 B. Malinowski – *Argonauts of the Western Pacific.* E. P. Dutton &
Co. Inc.: New York, 1922.

26 V. R. Padgett & D. O. Jorgenson – 'Superstition and economic
threat: Germany, 1918–1940', *Personality and Social Psychology
Bulletin* #8, pages 736–74. 1982.

27 G. Keinan – 'Effects of stress and tolerance of ambiguity on magical
thinking', *Journal of Personality and Social Psychology* #67(1),
pages 48–55. 1994.

28 C. Nemeroff & P. Rozin – 'The contagion concept in adult thinking
in the United States: Transmission of germs and interpersonal
influence', *Ethos* #22, pages 158–86. 1994.

  Related work is described in the article by P. Rozin, L. Millman
& C. Nemeroff – 'Operation of the laws of sympathetic magic in
disgust and other domains', *Journal of Personality and Social
Psychology* #50, pages 703–12. 1986.

29 J. Henry – 'Coincidence experience survey', *Journal of the Society
for Psychical Research*, #59(831), pages 97–108. 1993.

30 S. Milgram – *Obedience to authority: An experimental view.* Harper
& Row: New York, 1974.

31 C. L. Sheridan & R. G. King Jr. – 'Obedience to authority with an
authentic victim', *Proceedings of the 80th Annual Convention of the
American Psychological Association*, pages 165–6. 1972.

32 Described in T. Blass – *The man who shocked the world: The life
and legacy of Stanley Milgram.* Basic Books: New York, 2004.

33 S. Milgram – 'The small-world problem', *Psychology Today* #1,
pages 61–7. 1967.

— J. Travers & S. Milgram – 'An experimental study of the small
world problem', *Sociometry* #32, pages 425–43. 1969.

34 D. Watts – *Small Worlds: The Dynamics of Networks Between
Order and Randomness.* Princeton University Press: Princeton,
1999.

35 J. A. Paulos – *A Mathematician Reads the Newspaper.* Penguin
Books: London, 1995.

36 R. Wiseman – 'It really is a small world that we live in', *Daily Telegraph*, page 16, 4 June 2003.

37 D. Derbyshire – 'Physics too hot for a fire walker's feat', *Daily Telegraph*, page 17, 23 March 2000.

38 R. Wiseman, C. Watt, P. Stevens, E. Greening & C. O'Keeffe – 'An investigation into alleged "hauntings"', *The British Journal of Psychology* #94, pages 195–211. 2003.

39 R. Lange & J. Houran – 'Context-induced paranormal experiences: Support for Houran and Lange's model of haunting phenomena', *Perceptual and Motor Skills* #84, pages 1455–8. 1997.

40 V. Tandy & T. Lawrence – 'The ghost in the machine', *Journal of the Society for Psychical Research* #62, pages 360–4. 1998.

41 S. Angliss, GéNIA, C. O'Keeffe, R. Wiseman & R. Lord – 'Soundless music', in B. Arends & D. Thackara (eds), *Experiments: Conversations in art and science*, pages 139–71; quote from page 152. The Wellcome Trust: London, 2003.

42 A. Watson & D. Keating – 'Architecture and sound: An acoustic analysis of megalithic monuments in prehistoric Britain', *Antiquity* #73, pages 325–36. 1999.

43 P. Devereux – *Stone Age Soundtracks: The Acoustic Archaeology of Ancient Sites*. Vega Books: London, 2002.

4. Making your mind up

1 T. E. Moore – 'Subliminal perception: Facts and fallacies', *Skeptical Inquirer* #16, pages 273–81. 1992.

— A. Pratkanis – 'The cargo cult science of subliminal persuasion', *Skeptical Inquirer* #16, pages 260–72. 1992.

2 S. A. Lowery & M. L. DeFleur – *Milestones in Mass Communication Research: Media Effects* (3rd ed.). The information in this section is from the chapter 'Project Revere: Leaflets as a Medium of Last Resort'. Longman: White Plains, 1995.

3 M. L. DeFleur & R. M. Petranoff – 'A televised test of subliminal persuasion', *Public Opinion Quarterly* #23, pages 168–80. 1959.

4 B. Beyerstein – 'Subliminal self-help tapes: Promises, promises . . .', *Rational Enquirer* #6(1), 1993.

5 E. Eich & R. Hyman – 'Subliminal self-help' in D. Druckman &
   R. Bjork (eds), *In the Mind's Eye: Enhancing Human Performance*,
   pages 107–19. National Academy Press: Washington DC, 1991.
6 P. M. Merikle & H. Skanes – 'Subliminal self-help audio tapes:
   A search for placebo effects', *Journal of Applied Psychology* #77,
   pages 772–6. 1992.
7 S. Lenz – 'The effect of subliminal auditory stimuli on academic
   learning and motor skills performance among police recruits',
   unpublished doctoral dissertation, California School of Professional
   Psychology, Los Angeles, California, 1989.
8 B. Buchanan & J. L. Bruning – 'Connotative meanings of first names
   and nicknames on three dimensions', *Journal of Social Psychology*
   #85, pages 143–4. 1971.
9 A. A. Hartman, R. C. Nicolay & J. Hurley – 'Unique personal
   names as a social adjustment factor', *Journal of Social Psychology*
   #75, pages 107–10. 1968. Heldref Publications. Reprinted with
   permission.
10 H. Harari & J. W. McDavid – 'Name stereotypes and teachers'
   expectations', *Journal of Educational Psychology* #65, pages 222–5.
   1973.
11 W. F. Murphy – 'A note on the significance of names',
   Psychoanalytical Quarterly #26, pages 91–106. 1957.
12 N. Christenfeld, D. P. Phillips & L. M. Glynn – 'What's in a name:
   Mortality and the power of symbols', *Journal of Psychosomatic
   Research*, #47(3), pages 241–54. 1999.
13 G. Smith & S. Morrison – 'Monogrammic Determinism?',
   Psychosomatic Medicine #67, pages 820–4. 2005.
14 R. L. Zweigenhaft – 'The other side of unusual names', *Journal of
   Social Psychology* #103, pages 291–302. 1997. Heldref Publications.
   Reprinted with permission.
15 B. W. Pelham, M. C. Mirenberg & J. K. Jones – 'Why Susie sells
   seashells by the seashore: Implicit egotism and major life decisions',
   Journal of Personality and Social Psychology #82, pages 469–87.
   2002.
16 J. T. Jones, B. W. Pelham, M. Carvallo & M. C. Mirenberg –
   'How do I love thee? Let me count the Js: Implicit egotism and
   interpersonal attraction', *Journal of Personality and Social
   Psychology* #87(5), pages 655–83. 2004.

17 L. Casler – 'Put the blame on name', *Psychological Reports* #36, pages 467–72. 1975.

18 J. A. Bargh, M. Chen & L. Burrows – 'Automaticity of social behavior: Direct effects of trait construct and stereotype priming on action', *Journal of Personality and Social Psychology* #71, pages 230–44. 1996.

19 A. Dijksterhuis & A. van Knippenberg – 'The relation between perception and behavior, or how to win a game of Trivial Pursuit', *Journal of Personality and Social Psychology* #74(4), pages 865–77. 1998.

20 N. Gueguen – 'The effects of a joke on tipping when it is delivered at the same time as the bill', *Journal of Applied Social Psychology* #32, pages 1955–63. 2002.

21 N. Gueguen & P. Legoherel – 'Effect on tipping of barman drawing a sun on the bottom of customers' checks', *Psychological Reports* #87, pages 223–6. 2000.

— K. L. Tidd & J. S. Lockard – 'Monetary significance of the affiliative smile: A case for reciprocal altruism', *Bulletin of the Psychonomic Society* #11, pages 344–6. 1978.

— B. Rind & P. Bordia – 'Effect of server's "Thank You" and personalization on restaurant tipping', *Journal of Applied Social Psychology*, #25(9), pages 745–51. 1995.

22 M. R. Cunningham – 'Weather, mood, and helping behavior: Quasi experiments with the sunshine Samaritan', *Journal of Personality and Social Psychology* #37, pages 1947–56. 1979.

— B. Rind & D. Strohmetz – 'Effects of beliefs about future weather conditions on tipping', *Journal of Applied Social Psychology* #31(2), pages 2160–4. 2001.

23 K. Garrity & D. Degelman – 'Effect of server introduction on restaurant tipping', *Journal of Applied Social Psychology* #20, pages 168–72. 1990.

— K. M. Rodrigue – 'Tipping tips: The effects of personalization on restaurant gratuity', Master's thesis, Division of Psychology and Special Education, Emporia State University, 1999.

24 A. H. Crusco & C. G. Wetzel – 'The Midas Touch: The effects of interpersonal touch on restaurant tipping', *Personality and Social Psychology Bulletin* #10, pages 512–17. 1984.

25 C. S. Areni & D. Kim – 'The influence of background music on

shopping behavior: Classical versus top-forty music in a wine store', *Advances in Consumer Research* #20, pages 336–40. 1993.

26 J. N. Rogers – *The Country Music Message*. University of Arkansas Press, 1989.

27 S. Stack & J. Gundlach – 'The effect of country music on suicide', *Social Forces* #71(1), pages 211–18. 1992.

28 This controversial finding has been discussed in the following papers:

E. R. Maguire & J. B. Snipes – 'Reassessing the link between country music and suicide', *Social Forces* #72(4), pages 1239–43. 1994.

S. Stack & J. Gundlach – 'Country music and suicide: A reply to Maguire and Snipes', *Social Forces* #72(4), pages 1245–8. 1994.

G. W. Mauk, M. J. Taylor, K. R. White & T. S. Allen – 'Comments on Stack and Gundlach's "The effect of country music on suicide": An "achy breaky heart" may not kill you', *Social Forces* #72(4), pages 1249–55. 1994.

S. Stack & J. Gundlach – 'Psychological versus sociological perspectives on suicide: A reply to Mauk, Taylor, White, and Allen', *Social Forces* #72(4), pages 1257–61. 1994.

J. B. Snipes & E. R. Maguire – 'Country music, suicide, and spuriousness', *Social Forces* #74, pages 327–9. 1995.

S. Stack & J. Gundlach – 'Country Music and Suicide – Individual, Indirect, and Interaction Effects: A Reply to Snipes and Maguire', *Social Forces* #74(1), pages 331–5. 1995.

29 D. P. Phillips – 'The influence of suggestion on suicide: Substantive and theoretical implications of the Werther Effect', *American Sociological Review* #39, pages 340–54. 1974.

— id. – 'Motor vehicle fatalities increase just after a publicised suicide story', *Science* #196, pages 1464–5. 1977.

— id. – 'Airplane accident fatalities increase just after newspaper stories about murder and suicide', *Science* #201, pages 748–50. 1978.

— id. – 'Suicide, motor vehicle fatalities, and the mass media: Evidence towards a theory of suggestion', *American Journal of Sociology* #84, pages 1150–74. 1979.

— id. – 'Airplane accidents, murder, and the mass media: Towards a theory of imitation and suggestion', *Social Forces* #58(4), pages 1000–24. 1980.

— id. – 'The impact of fictional television stories on US adult fatalities: New evidence on the effect of the mass media on violence', *American Journal of Sociology* #87(6), pages 1340–59. 1982.

— id. – 'The impact of mass media violence on US homicides', *American Sociological Review* #48, pages 560–8. 1983.

30 S. Stack – 'Media coverage as a risk factor in suicide', *Journal of Epidemiology and Community Health* #57, pages 238–40. 2003.

31 L. F. Martel & H. B. Biller – *Stature and Stigma*. Lexington Books: Lexington, 1987.

32 B. Pawlowski, R. I. Dunbar & A. Lipowicz – 'Tall men have more reproductive success', *Nature* #403(6766), page 156. 2000.

33 T. Gregor – *The Mehinaku: The Dream of Daily Life in a Brazilian Indian Village*. University of Chicago Press: Chicago, 1977.

34 T. A. Judge & D. M. Cable – 'Effect of physical height on workplace success and income: Preliminary test of a theoretical model', *Journal of Applied Psychology* #89(3), pages 428–41. 2004.

35 P. R. Wilson – 'Perceptual distortion of height as a function of ascribed academic status', *Journal of Social Psychology* #74, pages 97–102. 1968.

36 H. H. Kassarjian – 'Voting intentions and political perception', *Journal of Psychology* #56, pages 85–8. 1963.

37 P. A. Higham & W. D. Carment – 'The rise and fall of politicians: The judged heights of Broadbent, Mulroney and Turner before and after the 1988 Canadian federal election', *Canadian Journal of Behavioral Science* #24, pages 404–9. 1992.

38 R. Highfield – 'Politicians: this is how we see them', *Daily Telegraph*, pages 22–3. 21 March 2001.

— R. Wiseman – 'A short history of stature', *Daily Telegraph*, page 22. 21 March 2001.

39 R. J. Pellegrini – 'Impressions of the male personality as a function of beardedness', *Psychology* #10, pages 29–33. 1973.

40 A. Todorov, A. N. Mandisodza, A. Goren & C. C. Hall – 'Inferences of competence from faces predict election outcomes', *Science* #308, pages 1623–6. 2005.

41 J. E. Stewart II – 'Defendants' attractiveness as a factor in the outcome of trials', *Journal of Applied Social Psychology* #10, pages 348–61. 1980.

42 R. B. Cialdini – *Influence: Science and Practice*. Allyn & Bacon: Needham Heights, MA, 2001.

43 S. M. Smith, W. D. McIntosh & D. G. Bazzini – 'Are the beautiful good in Hollywood? An analysis of stereotypes on film', *Basic and Applied Social Psychology* #21, pages 69–81. 1999.

44 D. G. Dutton & A. P. Aron – 'Some evidence for heightened sexual attraction under conditions of high anxiety', *Journal of Personality and Social Psychology* #30, pages 510–17. 1974.

45 C. Bale, R. Morrison & P. G. Caryl – 'Chat up lines as male sexual displays', *Personality and Individual Differences* #40, pages 655–64. 2006.

46 B. Fraley & A. Aron – 'The effect of a shared humorous experience on closeness in initial encounters', *Personal Relationships* #11, pages 61–78. 2004.

47 J. E. Smith & V. A. Waldorf – '"Single white male looking for thin, very attractive . . ."', *Sex Roles* #23, pages 675–85. 1990.

5. The scientific search for the world's funniest joke

1 H. R. Pollio & J. W. Edgerly – 'Comedians and comic style', in A. J. Chapman & H. C. Foot (eds) *Humor and Laughter: Theory, Research, and Applications*. pages 215–44. Transaction: New Jersey, 1996.

2 C. Davies – 'Jewish jokes, anti-Semitic jokes and Hebredonian jokes', in A. Ziv (ed.), *Jewish Humour*, pages 59–80. Papyrus Publishing House: Tel Aviv, 1986.

3 H. A. Wolff, C. E. Smith & H. A. Murray – 'The Psychology of humor: 1. A study of responses to race-disparagement jokes', *Journal of Abnormal and Social Psychology* #28, pages 345–65. 1934.

4 J. Morreall – *Taking Laughter Seriously*. State University of New York Press: Albany, 1983.

5 G. R. Maio, J. M. Olson, & J. Bush – 'Telling jokes that disparage social groups: Effects on the joke teller's stereotypes', *Journal of Applied and Social Psychology* #27(22), pages 1986–2000. 1997.

6 B. Seibt & J. Förster – 'Risky and careful processing under stereotype threat: How regulatory focus can enhance and deteriorate

Notes

performance when self stereotypes are active', *Journal of Personality and Social Psychology* #87, pages 38–56. 2004.

7 R. Provine – *Laughter: A Scientific Investigation.* Viking: New York, 2000.

8 M. Middleton & J. Moland – 'Humor in Negro and White subcultures: A study of jokes among university students', *American Sociological Review* #24, pages 61–9. 1959.

9 P. J. Castell & J. H. Goldstein – 'Social occasions of joking: A cross cultural study' in A. J. Chapman & H. C. Foot (eds), *It's a Funny Thing, Humour*, pages 193–7. Pergamon Press: Oxford, 1976.

10 J. B. Levine – 'The feminine routine', *Journal of Communication* #26, pages 173–5. 1976.

11 L. La Fave, J. Haddad & W. A. Maesen – 'Superiority, enhanced self-esteem, and perceived incongruity humour theory' in T. Chapman & H. Foot (eds), *Humor and Laughter: Theory, Research and Applications*, pages 63–91. Transaction: New Jersey. Copyright c. 1996 by Transaction. Reprinted with permission of the publisher.

12 Sir Harry Kroto originally presented us with a version of the joke in broad Glaswegian. As we knew that the material in *LaughLab* would be read by people across the world, we created a version that would allow a much larger number of people to appreciate the joke. Sir Harry Kroto's original entry, which he much prefers, is reproduced below.

A guy is walking along the road in Glasgow and sees a man with a humungous great dog on the other side of the street. He goes over and says, 'Hey Jimmy, dis yer dawg byte?'

The man says, 'Nu.'

So the guy pats the dog on the head, whereupon the dog snaps, and bites off a couple of fingers. 'Grrrrwrwrwrwrrfraarrrrrrrrrrggggggggklle . . . umph.

The guy screams, 'Aaaghgee,' as blood streams from his hand, and shouts, 'A tawt yer said yer dawg dusna byte.'

The man says quietly with a look of calm diffidence, 'Sna ma dawg.'

13 K. Binsted & G. Ritchie – 'Computational rules for punning riddles', *Humor: International Journal of Humor Research* #10(1), pages 25–76. 1997.

14 M. Le Page – 'Women's orgasms are a turn-off for the brain', *New Scientist*, page 14, 25 June 2005.

15 H. H. Brownell & H. Hardner – 'Neuropsychological insights into humour' in J. Durant and J. Miller (eds), *Laughing Matters: A Serious Look at Humour*, pages 17–35. Longman: Harlow, 1988.

16 T. Friend – 'What's so funny?' *New Yorker*, pages 78–93. 11 November 2002.

17 D. Barry – 'Send in your weasel jokes', *International Herald Tribune*. 19–20 January 2002.

18 id. – *Dave Barry Talks Back*. See chapter entitled 'Introducing: Mr Humor Person'. Crown Publishers, Inc.: New York, 1991.

19 F. Strack – 'Inhibiting and facilitating conditions of the human smile: A nonobtrusive test of the facial feedback hypothesis', *Journal of Personality and Social Psychology* #54(5), pages 768–77. 1988.

20 V. B. Hinsz & J. A. Tomhave – 'Smile and (half) the world smiles with you, frown and you frown alone', *Personality and Social Psychology Bulletin* #17(5), pages 586–92. 1991.

21 A. M. Rankin & P. J. Philip – 'An epidemic of laughing in the Bukoba District of Tanganyika', *Central African Journal of Medicine* #9, pages 167–70. 1963.

22 S. S. Janus – 'The great comedians: Personality and other factors', *The American Journal of Psychoanalysis* #35, pages 169–74. 1975. Quote reproduced with kind permission of Springer Science and Business Media.

23 S. Fisher & R. L. Fisher – *Pretend the World is Funny and Forever: A psychological analysis of comedians, clowns, and actors*. Lawrence Erlbaum Associates: Hillsdale, 1981.

24 J. Rotton – 'Trait humor and longevity: Do comics have the last laugh?' *Health Psychology* #11(4), pages 262–6. 1992.

25 H. M. Lefcourt – 'Humor', in C. R. Snyder & S. J. Lopez (eds), *Handbook of Positive Psychology*, pages 619–31. Oxford University Press: Oxford, 2005.

26 H. Lefcourt, K. Davidson-Katz & K. Kueneman – 'Humor and immune system functioning', *International Journal of Humor Research* #3, pages 305–21. 1990.

27 J. Rotton & M. Shats – 'Effects of state humor, expectancies, and choice on postsurgical mood and self-medication: A field

experiment', *Journal of Applied Social Psychology* #26, pages 1775–94. 1996.

28 H. M. Lefcourt, K. Davidson, R. Shepherd, M. Phillips, K. Prkachin & D. Mills – 'Perspective-taking humor: Accounting for stress moderation', *Journal of Social and Clinical Psychology* #14, pages 373–91. 1995.

29 D. Keltner & G. A. Bonanno – 'A study of laughter and dissociation: Distinct correlates of laughter and smiling during bereavement', *Journal of Personality and Social Psychology* #73, pages 687–702. 1997.

— G. A. Bonanno & D. Keltner – 'Facial expressions of emotion and the course of conjugal bereavement', *Journal of Abnormal Psychology* #106, pages 126–37. 1997.

30 V. Saroglou – 'Sense of humor and religion: An a priori incompatibility? Theoretical considerations from a psychological perspective', *Humor: International Journal of Humor Research* #15, pages 191–214. 2002.

31 id. – 'Religiousness, religious fundamentalism, and quest as predictors of humor creation', *International Journal for the Psychology of Religion* #12, pages 177–88. 2002.

32 V. Saroglou & J. M. Jaspard – 'Does religion affect humour creation? An experimental study'. *Mental Health, Religion, and Culture* #4, pages 33–46. 2001.

33 H. J. Eysenck – 'National differences in "sense of humor": Three experimental and statistical studies', *Journal of Personality* #13(1), pages 37–54. 1944.

34 T. Radford – 'Don't gag on it, but this is what has us all in stitches', *Guardian*, page 6. 4 October 2003.

6. Sinner or saint?

1 R. T. LaPiere – 'Attitudes versus actions', *Social Forces* #13(2), pages 230–7. 1934.

2 J. Trinkaus – 'Color preference in sport shoes: An informal look', *Perceptual and Motor Skills* #73, pages 613–14. 1991.

3 id. – 'Television station weather-persons' winter storm predictions:

An informal look', *Perceptual and Motor Skills* #79, pages 65–6. 1994.

4 id. – 'Wearing baseball-type caps: An informal look', *Psychological Reports* #74(2), pages 585–6. 1994.

5 id. – 'The demise of "Yes": An informal look', *Perceptual and Motor Skills* #84, page 866. 1997.

6 id. – 'Preconditioning an audience for mental magic: An informal look', *Perceptual and Motor Skills* #51, page 262. 1980.

7 id. – 'The attaché case combination lock. An informal look', *Perceptual and Motor Skills* #72, page 466. 1991.

8 R. P. Feynman – *Surely You're Joking, Mr Feynman!*. Random House: London, 1992.

9 J. Trinkaus – 'Gloves as vanishing personal "stuff": An informal look', *Psychological Reports* #84, pages 1187–8. 1999.

10 M. S. C. Lim, M. E. Hellard & C. K. Aitken – 'The case of the disappearing teaspoons: Longitudinal cohort study of the displacement of teaspoons in an Australian research institute', *British Medical Journal* #331, pages 1498–1500. 2005.

11 B. Herer – 'Disappearing teaspoons', *British Medical Journal* #332, page 121. 2006.

12 J. Trinkaus – 'Compliance with the item limit of the food supermarket express checkout lane: An informal look', *Psychological Reports* #73, pages 105–6. 1993.

13 id. – 'Compliance with the item limit of the food supermarket express checkout lane: Another look', *Psychological Reports* #91, pages 1057–8. 2002.

14 id. – 'School zone limit dissenters: An informal look', *Perceptual and Motor Skills* #88, pages 1057–8. 1999.

15 id. – 'Stop sign compliance: An informal look', *Psychological Reports* #89, pages 1193–4. 1999.

16 id. – 'Blocking the box: An informal look', *Psychological Reports* #89, pages 315–16. 2001.

17 id. – 'Shopping center fire zone parking violators: An informal look', *Perceptual and Motor Skills* #95, pages 1215–16. 2002.

18 US News – 'Oprah: A heavenly body? Survey finds talk-show host a celestial shoo-in', page 18. 31 March 1997.

19 S. A. Hellweg, M. Pfau & S. B. Brydon – '*Televised presidential*

Notes

Notes

Notes

Notes

Notes

Notes

Notes

Notes

Notes

Notes

Notes

Notes

Notes

Notes

Notes

Notes

Notes

Notes

Notes

Notes

Notes

20 P. Jaret – 'Blinking and thinking', *In Health* #4(4), pages 36–7. 1990.

21 J. J. Tecce – 'Body language in Presidential debates as a predictor of election results: 1960–2004', unpublished report, Boston College, 2004.

22 P. Suedfeld, S. Bochner & D. Wnek – 'Helper–sufferer similarity and a specific request for help: Bystander intervention during a peace demonstration', *Journal of Applied Social Psychology* #2, pages 17–23. 1972.

23 J. P. Forgas – 'An unobtrusive study of reactions to national stereotypes in four European countries', *Journal of Social Psychology* #99, pages 37–42. 1976.

24 A. N. Doob & A. E. Gross – 'Status of frustrator as an inhibitor of horn-honking responses', *Journal of Social Psychology* #76, pages 213–18. 1968.

25 F. K. Heussenstamm – 'Bumper stickers and the cops', *Transaction* #8, pages 32–3. 1971.

26 J. M. Burger, N. Messian, S. Patel, A. del Prado & C. Anderson – 'What a coincidence! The effects of incidental similarity on compliance', *Personality and Social Psychology Bulletin* #30, pages 35–43. 2004.

27 J. F. Finch & R. B. Cialdini – 'Another indirect tactic of (self-)image management', *Personality and Social Psychology Bulletin* #15, pages 222–32. 1989.

28 S. Milgram & R. L. Shotland – *Television and Antisocial Behaviour: Field Experiments.* Academic Press: New York, 1973.

29 Both of these surveys are cited in Milgram & Shotland 1973.

30 A. Huston, E. Donnerstein, H. Fairchild, N. D. Feshbach, P. A. Katz, J. P. Murray, E. A. Rubinstein, B. L. Wilcox & D. Zuckerman – *Big World, Small Screen: The Role of Television in American Society.* University of Nebraska Press: Lincoln, 1992.

31 S. Milgram – 'The lost-letter technique', *Psychology Today* #3(3), pages 32–3, 66, 68. 1969.

32 F. S. Bridges & P. C. Thompson – 'Impeachment affiliation and levels of response to lost letters', *Psychological Reports* #84, pages 828–31. 1999.

297

33 B. J. Bushman & A. M. Bonacci – 'You've got mail: Using e-mail to examine the effect of prejudiced attitudes on discrimination against Arabs', *Journal of Experimental Social Psychology* #40, pages 753–9. 2004.

34 V. Saroglou, I. Pichon, L. Trompette, M. Verschueren & R. Dernelle – 'Prosocial behavior and religion: New evidence based on projective measures and peer ratings', *Journal for the Scientific Study of Religion* #44, pages 323–48. 2005.

35 Quote reproduced with permission of authors and publisher from: G. B. Forbes, K. T. Vault & H. F. Gromoll – 'Willingness to help strangers as a function of liberal, conservative or catholic church membership: A field study with the lost-letter technique', *Psychological Reports* #28, pages 947–9. © Psychological Reports 1971.

36 J. M. Darley & C. D. Batson – '"From Jerusalem to Jericho": A study of situational and dispositional variables in helping behavior', *Journal of Personality and Social Psychology* #27, pages 100–8 1973.

37 R. Levine, T. Martinez, G. Brase & K. Sorenson – 'Helping in 36 US cities', *Journal of Personality and Social Psychology* #67, pages 69–81. 1994.

38 R. V. Levine, A. Norenzayan & K. Philbrick – 'Cross-cultural differences in helping strangers', *Journal of Cross-Cultural Psychology* #32, pages 543–60. 2001.

39 S. Milgram – 'The experience of living in cities', *Science* #167, pages 1461–8. 1970.

40 R. V. Levine & A. Norenzayan – 'The pace of life in 31 countries', *Journal of Cross-Cultural Psychology* #30, pages 178–205. 1999.

41 R. V. Levine, K. Lynch, K. Miyake & M. Lucia – 'The type A city: Coronary heart disease and the pace of life', *Journal of Behavioral Medicine* #12, pages 509–24. 1989.

42 J. A. M. Farver, B. Welles-Nyström, D. L. Frosch, S. Wimbarti & S. Hoppe-Graff – 'Toy Stories: Aggression in Children's Narratives in the United States, Sweden, Germany, and Indonesia', *Journal of Cross-Cultural Psychology* #28(4), pages 393–420. 1997.

43 P. G. Zimbardo – 'The human choice: Individuation, reason, and order versus deindividuation, impulse, and chaos' in W. J. Arnold

and D. Levine (eds), *1969 Nebraska Symposium on Motivation*, pages 237–307. University of Nebraska Press: Lincoln, NE, 1970.

44 id. – 'Foreword', in S. Milgram, J. Sabini, and M. Silver (eds), *The Individual in the Social World: Essays and Experiments, 2nd Edition*, pages ix–xi. McGraw-Hill: New York, 1992.

45 J. L. Freedman & S. C. Fraser – 'Compliance without pressure: The foot-in-the-door technique', *Journal of Personality and Social Psychology* #4, pages 196–202. 1966.

Epilogue

1 Due to logistical issues, the measurements in New York and Croatia were taken between 11.30 a.m. and 2 p.m. on 1 October 2006, and 5 September 2006, respectively.

2 The mean walking times (in seconds) were as follows: Singapore, 10.55; Copenhagen, 10.82; Madrid, 10.89; Guangzhou, 10.94; Dublin, 11.03; Curitiba, 11.13; Berlin, 11.16; New York, 12.00; Utrecht, 12.04; Vienna, 12.06; Warsaw, 12.07; London, 12.17; Zagreb, 12.20; Prague, 12.35; Wellington, 12.62; Paris, 12.65; Stockholm, 12.75; Ljubljana, 12.76; Tokyo, 12.83; Ottawa, 13.72, Harare, 13.92; Sofia, 13.96; Taipei, 14.00; Cairo, 14.18; Sana'a, 14.29; Bucharest, 14.36; Dubai, 14.64; Damascus, 14.94; Amman, 15.95; Bern, 17.37; Manama, 17.69; Blantyre, 31.60.

Towards the Worldwide Eradication of 'FTSE-ITIS'

1 I made this bit up.